THE ORIGINS OF HUMAN BEHAVIOUR

TITLES OF RELATED INTEREST

THE ORIGINS OF
HUMAN BEHAVIOUR

Edited by R. A. Foley

Department of Biological Anthropology, University of Cambridge

Routledge

Taylor & Francis Group

LONDON AND NEW YORK

First published in 1991
Reprinted 2000
by Routledge
2 Park Square, Milton Park, Abingdon, Oxfordshire OX14 4RN
711 Third Avenue, New York, NY 10017

Transferred to Digital Printing 2006

First issued in paperback 2014

Routledge is an imprint of the Taylor & Francis Group, an informa business

---

British Library Cataloguing in Publication Data

The origins of human behaviour. – (One world archaeology)
1. Man. Behaviour. Origins
I. Foley, Robert II. Series
155.7

ISBN 978-0-415-75491-0 (pbk)

Library of Congress Cataloging in Publication Data

The Origins of human behaviour/edited by R. A. Foley
    p.   cm. – (One world archaeology: 19)
    Includes bibliographical references and index
    ISBN 978-0-044-45015-3 (hbk)
    ISBN 978-0-415-75491-0 (pbk)
    1. Human evolution.   2. Human behaviour – Origin.
3. Behavior evolution. I. Foley, Robert.   II. Series.
GN281.4.075   1991
573.2–dc20   90–12920 CIP

Typeset in 10 on 11 point Bembo
by Computape (Pickering) Ltd, North Yorkshire

Publisher's Note
The publisher has gone to great lengths to ensure the quality of this reprint
but points out that some imperfections in the original may be apparent

List of contributors

R. A. Foley, Department of Biological Anthropology, University of Cambridge, UK.
Phillip J. Habgood, Department of Anthropology, University of Sydney, NSW, Australia.
Sheldon Klein, Computer Sciences Department, University of Wisconsin-Madison, USA.
W. C. McGrew, Department of Psychology, University of Stirling, UK.
Francis B. Musonda, Livingstone Museum, Livingstone, Zambia.
Thomas Wynn, Department of Anthropology, University of Colorado at Colorado Springs, USA.
Ezra B. W. Zubrow, Department of Anthropology, University of New York at Buffalo. USA.

Foreword

This book is one of a major series of more than 20 volumes resulting from the World Archaeological Congress held in Southampton, England, in September 1986. The series reflects the enormous academic impact of the Congress, which was attended by 850 people from more than 70 countries, and attracted many additional contributions from others who were unable to attend in person.

The *One World Archaeology* series is the result of a determined and highly successful attempt to bring together for the first time not only archaeologists and anthropologists from many different parts of the world, as well as academics from a host of contingent disciplines, but also non-academics from a wide range of cultural backgrounds, who could lend their own expertise to the discussions at the Congress. Many of the latter, accustomed to being treated as the 'subjects' of archaeological and anthropological observation, had never before been admitted as equal participants in the discussion of their own (cultural) past or present, with their own particularly vital contribution to make towards global, cross-cultural understanding.

The Congress therefore really addressed world archaeology in its widest sense. Central to a world archaeological approach is the investigation not only of how people lived in the past but also of how, and why, changes took place resulting in the forms of society and culture which exist today. Contrary to popular belief, and the archaeology of some 20 years ago, world archaeology is much more than the mere recording of specific historic events, embracing as it does the study of social and cultural change in its entirety. All the books in the *One World Archaeology* series are the result of meetings and discussions which took place within a context that encouraged a feeling of self-criticism and humility in the participants about their own interpretations and concepts of the past. Many participants experienced a new self-awareness, as well as a degree of awe about past and present human endeavours, all of which are reflected in this unique series.

The Congress was organized around major themes. Several of these themes were based on the discussion of full-length papers which had been circulated some months previously to all who had indicated a special interest in them. Other sessions, including some dealing with areas of specialization defined by period or geographical region, were based on oral addresses, or a combination of precirculated papers and lectures. In all cases, the entire sessions were recorded on cassette, and all contributors were presented with the recordings of the discussion of their papers. A major part of the thinking behind the Congress was that such a meeting of many hundreds of participants that did not leave behind a published record of its academic discussions would be little more than an exercise in tourism.

Thus, from the very beginning of the detailed planning for the World

Archaeological Congress, in 1982, the intention was to produce post-Congress books containing a selection only of the contributions, revised in the light of discussions during the sessions themselves as well as during subsequent consultations with the academic editors appointed for each book. Particularly in the case of sessions based on precirculated papers, all contributors were aware of the subsequent publication production schedules – if their papers were selected for publication they would have only a few months to revise them according to editorial specifications, and they would become authors in an important academic volume scheduled to appear within a reasonable period following the Southampton meeting.

The origins of human behaviour results from the four and a half days at the Congress of discussions of precirculated papers and verbal contributions presented under the overall title 'The Pleistocene Perspective', organized by Professor Michael Day, Dr Robert Foley and Mr Arthur ApSimon. More specifically, it derives from the Congress subtheme on 'Hominid Behaviour and Ecology', which was organized by the editor of this book. Other contributions originally linked to this overall theme of the Congress have been published in *Hunters of the recent past* (edited by L. B. Davis & B. O. K. Reeves), *The world at 18 000 BP: low latitudes* (edited by C. Gamble & O. Soffer) and *The world at 18 000 BP: high latitudes* (edited by O. Soffer & C. Gamble).

Unfortunately the publication of this book has been much delayed and is the last of those in this series to concentrate on the Pleistocene (those on 'Human Evolution', 'Hominid Dispersal Patterns' and 'Adaptations at around the Pleistocene/Holocene Boundary' having, very reluctantly, been laid aside). The publication of *The origins of human behaviour* reflects, therefore, the commitment by the World Archaeological Congress to palaeoanthropology and palaeolithic archaeology despite the very complex and troubled history of the organization of sessions on 'The Pleistocene Perspective' at the 1986 Congress – reviewed in detail elsewhere (Ucko 1987, pp. 24, 34, 39–40, 61, 64, 82, 125, 129, 131–2, 144, 151, 214, 222, 224, 235, 247, 249, 266–9) – which inevitably led to a lack of thematic coherence, and a somewhat variable quality, in the papers presented at that time. Since then, numerous attempts have been made to create several thematic books on the Pleistocene from the many 1986 contributions but, finally, it has been decided to concentrate exclusively on the present book.

The origins of human behaviour is much more than simply a symbolic publication. Its concerns are central to many of the overall perceptions of the *One World Archaeology* series, focusing attention on such very basic questions as what constitutes 'culture', whether such a term is merely descriptive or whether it may be used as an analytical tool, and how it can – or cannot – serve to qualify the 'human condition'. As such, *The origins of human behaviour* is implicitly, and necessarily, concerned with concepts such as innovation, discovery and diffusion – making it an interesting companion to *What's new? A closer look at the process of innovation* (edited by S. E. van der Leeuw & R. Torrence) – and it challenges the reader to determine whether there is any reason against assuming their equal importance to the non-human sphere.

Likewise, this book also has much of importance to say about the subject matter of three other *One World Archaeology* books: *Animals into art* (edited by H. Morphy), *Archaeological approaches to cultural identity* (edited by S. J. Shennan) and *What is an animal?* (edited by T. Ingold). With regard to the first, the reader is urged to consider in detail the implications for the evolution of human organization and human thought of assumptions about any regular use of 'rules' in the schemata of Upper Palaeolithic artworks. With regard to the second, the problems of using archaeological evidence to establish whether or not there really were two distinct forms of humans living contemporaneously, apart or together, should caution against too easy assumptions about ethnic visibility in the archaeological record. In the last case – if culture is not an exclusively 'human' phenomenon – what really are the distinguishing points (if any) between human and animal?

The nature of these kinds of enquiries are both wide-ranging and comparative; far removed from the climate of much prevalent opinion about later archaeological material which currently stresses the necessity of inward-looking, intracultural, analysis and interpretation. It is exciting, therefore, to find renewed interest in other generalizing comparative approaches to human activity, such as the possible relevance of Piagetian constructs to archaeological data and their interpretation.

This book therefore foreshadows other attempts (e.g. Ucko 1990) to reinvestigate the possible relevance to current archaeological theory of grand schemes – mainly constructed in the 19th century by psychologists as well as anthropologists and archaeologists – about the human past, all being based on what has often been referred to as 'the comparative method'. It is fascinating to discover from this book how the need to investigate the mechanisms of non-genetic generational transmission of behavioural characteristics remains of central concern to current interpretations of human behaviour in the Pleistocene. Classic 'recapitulation' theories may be long outdated in biological terms, but curiosity about an assumed similarity between the development from an infant to an adult and the growth of human 'cultural' complexity is still in evidence.

The origins of human behaviour is a book which challenges the archaeologist to reconsider the appropriateness of many interpretative models in use in this discipline for periods much later than the Pleistocene. It also, however, serves as a challenge to its own practitioners: in publicly stressing the common humanity of its data base, and therefore claiming an almost Olympian detachment from the political and social debates surrounding contemporary human rights, Pleistocene studies have remained very much the exclusive domain of specialists from, or influenced by, a restricted Western background. If wide-ranging comparative methodology is to be its *forte*, such comparative schemata cannot afford to be based on assumptions deriving exclusively either from ethology or from presumed linear progressions of social development. The challenge is to refrain from sheltering behind the chronological remoteness of the material under study and – as some (Cann *et al.* 1987) have begun – to bring it into the public domain, demonstrating its relevance to those living in the present.

P. J. Ucko
Southampton

References

Cann, R. L., Stoneking, M. & Wilson, A. C. 1987. Mitochondial DNA and human evolution. *Nature* **325**, 31–6.
Ucko, P. J. 1987. *Academic freedom and apartheid: the story of the World Archaeological Congress.* London: Duckworth.
Ucko, P. J. 1990. *Whose culture is it anyway?* Frazer Lecture, University of Glasgow.

Contents

Preface

The World Archaeological Congress meetings in Southampton in September 1986 included a series of sessions on the problems of Pleistocene archaeology. The chapters in this book derive from some of those discussions. While the original meetings were extremely diverse, this volume focuses on the problems that face prehistorians and palaeoanthropologists trying to understand the long-term evolution of human behaviour and the patterns observable in the fossil and archaeological record of a period of time stretching over several million years. It is not the intention to present a comprehensive analysis of the origins and evolution of modern human behaviour, but rather to illustrate the diversity of approaches and concepts that are required if we are to unravel what must surely be the most complex of problems facing archaeologists and evolutionary biologists.

Much gratitude is owed to the many people who contributed to the original discussions and made the meeting a memorable one, as well as to the organizers, especially Peter Ucko, without whose drive little would have been achieved. I would also like to express my appreciation to those who chaired sessions during the Congress: Paul Callow, Michael Day, Clive Gamble, Wu Rukang, Chris Stringer and Jiří Svoboda. Thanks must also go to Marta Lahr and Harriet Eeley for their helpful comments on many of the ideas discussed in this book.

R. A. Foley
Cambridge

THE ORIGINS OF HUMAN BEHAVIOUR

Introduction: investigating the origins of human behaviour

R. A. FOLEY

It is hard to find a branch of anthropology and archaeology as disparate as the study of the origins and evolution of human behaviour, which may be referred to as behavioural palaeoanthropology.

At the outset there is the question of scale. The closest living relatives of hominids are the African apes, and most probably the chimpanzee (*Pan troglodytes* and *Pan paniscus*). Current evidence suggests that the split from *Pan* occurred between 5 and 8 million years ago (Holmes *et al.* 1989). The differences between these two groups of hominoids therefore developed over a period of several million years, a timescale of unique length within anthropology (although relatively short on a general palaeontological and evolutionary scale). Such a timescale makes it very difficult to conceptualize processes over long periods, to explain long-term events, and to document the timing and sequence of the major developments. As a result, the question of whether the shift from archaic to modern hominids was of evolutionary significance in itself or whether the principal evolutionary changes occurred with the appearance of the genus *Homo* some 2 million years earlier (Foley 1989), for example, remains problematic.

Scale is not simply a chronological consideration. Variability is central to all science, and anthropology is no exception. However, most of anthropology (including prehistoric archaeology) is concerned only with the development of models to describe and theories to explain intraspecific variability. Palaeoanthropology, though, must treat not just intraspecific variability but also variations across species, genera, and even higher taxonomic levels, while at the same time employing where necessary techniques, concepts, and assumptions developed both for later periods and for cultural rather than biological variability in behaviour.

Next comes the difficulty of lines of evidence. As McGrew points out (Ch. 1), behaviour does not fossilize but must be inferred indirectly from either fossil hominid morphology or archaeological remains. Both the fossil and archaeological records are notoriously incomplete. This is true of both the frequency with which prehistoric hominids and their activities are preserved and discovered, and the range of such activities that are subject to fossilization processes. We have access to only a minuscule proportion of the hominids that have ever lived, and to only a limited range of the results of their activities, these being confined principally to subsistence and technology.

This has led to a recognition that supporting evidence must come from other, indirect, sources. In practice this means some sort of analogical reasoning based on living groups of humans or animals. But here again difficulties arise. What are appropriate analogues for events and processes in early hominid evolution? Do modern humans provide suitable material for drawing parallels? Traditionally, living hunter-gatherers have been used to 'flesh out the fossil record', but it is clear, as Musonda discusses (Ch. 3), that there are as many differences between these groups and earlier hominids as there are similarities: they are members of a different species, they have undergone their own subsequent evolutionary and historical developments, and they are the product of specific historical, cultural, and environmental contexts (McGrew 1990). Furthermore, their technology is significantly different from that of prehistoric, nonmodern human populations. Simplistic hunter-gatherer models have been heavily criticized for being based on outmoded notions of cultural evolution that place living hunter-gatherers in a primitive, ancestral state.

The alternative source of analogue models is the nonhuman primates, but many difficulties arise here as well. While we share a close phylogenetic relationship and evolutionary history with other primates, they too have evolved in response to their own, unique, evolutionary context. Apes and monkeys are not just hominids *manqués*; they are species as well adapted to their niches as the hominids were to theirs. Analogies may be based on either phylogenetic or environmental similarity, or both. Historically, chimpanzees and baboons respectively have been taken as models for the two roles, but again there are difficulties with this approach, because evolution is the result of interaction between phylogeny and environment, and for every species this produces a unique set of circumstances.

Even if such analogies are useful, the next question that arises is what exactly is it in hominid evolution that ones wishes to model and seek an explanation for. A wide variety of distinctive human features have been identified and proposed as crucial to the evolution of the hominids from a more primitive ape precursor. These include bipedalism, enlarged brains and intelligence, tool-making, food-sharing, language, consciousness, enlarged social networks, hunting, territoriality, and so on. Historically, there has been a shifting emphasis among palaeoanthropologists – for example, from brains and technology to bipedalism and subsistence – that reflects a variety of factors both internal and empirical and external and political. As a result a wide range of social, behavioural, and ecological traits have been selected as the key to the change from animal to human behaviour, and work on the origins of human behaviour has necessarily been spread very thinly, with little consensus about where the main issues lie or, perhaps more importantly, how the behavioural variables are related to each other. For example, is speech significant in the processes of encephalization that occurred during hominid evolution, and how does it relate, chronologically and functionally, to other attributes such as conscious thought, hunting behaviour, social cooperation, etc? While McGrew shows that the range, complexity and variation of chimpanzee behaviour are greater than previously thought (Ch. 1), nonetheless analogical approaches remain problematic.

Even taking into account these essentially empirical, methodological, and technical questions, there remains yet one more black box for the palaeoanthropologist to try to peer into: that of explanation. The study of human evolution, especially its behavioural aspects, is on the cusp of the social and natural sciences. On one side lie essentially Darwinian explanations and theories, couched in terms of natural selection, reproductive advantage, costs and benefits, and evolutionarily stable strategies – the terms and concepts of behavioural ecology and sociobiology. On the other side lies social and cultural anthropology, with its armoury of functionalism and structuralism, cultural ecology and cultural relativism, social theory and societal norms. In between lie approaches that have attempted to integrate evolutionary and cultural theories (Shennan 1989). The problem is reminiscent of a large and complex landscape divided by a deep and meandering river. It is that river that somewhere during the course of their evolution hominids are thought to have crossed. But when and how? Until these questions are answered, what sort of explanation should be sought for hominid behavioural evolution – Darwinian or cultural – will remain unclear. (In fact, at present there is not even any consensus that hominids ever crossed the river at all; some feel they are still firmly placed on the sociobiological banks!)

The study of the origins of human behaviour, then, is a difficult subject, beset by problems of temporal and taxonomic scale, imperfect and incomplete evidence, an uncertain comparative framework, and competing explanatory systems. Given all that, behavioural palaeoanthropology may be likened to Samuel Johnson's comments on bipedal dogs: 'It is not done well; but you are surprised to find it done at all.'

Great advances, though, have been made in turning a subject that is potentially an exercise in hindsight into an empirically testable project with its own procedures (Kinzey 1987). Three factors are most probably responsible. First, there has been a considerable improvement in the fossil record of hominid evolution. This statement is not simply a reiteration of the oft-heard claim that the jigsaw of prehistory is gradually being filled in but is based on one aspect of recent developments in palaeoanthropology. This is that as more fossils have come to light, it is clear that hominid evolution is not a simple unilinear process from the primitive to the advanced; rather, it is a complex process involving cladogenetic speciation, extinction, and coexistence of species. As a result of this single discovery it is no longer possible to be vague about phases of hominid evolution. As McGrew points out (Ch. 1), if we are to use chimpanzees as a model for earlier hominids, we should know for which hominid they are appropriate – common ancestor, australopithecine, early *Homo*. Behavioural palaeoanthropology must explain variability in terms other than the chronological.

Second, field studies of primates have provided a vast amount of information about the evolutionary and behavioural ecology of complex and highly social mammals. This has been vital in opening up discussion about early hominids beyond such vague issues as male dominance and large groups. Of particular importance has been the move away from descriptions of

stereotypic species behaviour towards the recognition that behaviour is variable and flexible within species, according to such factors as age, sex, rank, and
ecological conditions (Dunbar 1988, Standen & Foley 1989). This means that
application of these results to the early hominids can go beyond single species
models ('the baboon analogy') to deal with issues of individual reproductive
strategies and life history correlates.

The third and related factor is the development of evolutionary theory
itself. Of particular importance has been the increasing recognition of two
facts; that behaviour evolves through the mechanism of natural selection and
that behavioural evolution can play a significant part in determining overall
patterns of evolution (Bateson 1988, Shennan 1989). The integrating of
behavioural, especially social (Humphrey 1976), evolution within evolutionary biology as a whole has taken human evolution, with its special behavioural
and social considerations, away from the margins of the subject and into the
mainstream.

What is perhaps most interesting is that these critical developments have
come not from archaeology, theoretically the subject most directly concerned
with the behaviour of early hominids, but from the adjacent field of evolutionary biology. Certainly there have been critical developments both in
methodology and in empirical results within archaeology (and nowhere has
this been more significant than in studies of early hominid meat-eating
behaviour), but the key development has been the recognition that evolutionary biology has the power to explain extremely complex behavioural
phenomena without recourse to the 'special case' of humans that has dogged
research into our origins.

It is important to understand the background to the problems currently
facing those researching into the origins and evolution of human behaviour.
However, it is perhaps even more vital to determine the way forward. The
problem appears overwhelmingly insoluble. Questions of 'origins' by definition relate to events in the past, which can never be directly observed, and
therefore rest uneasily on the boundaries of practical scientific investigation. If
science is the art of the soluble, as Peter Medawar has claimed (1967), it may be
that all discussions of human origins must remain speculative, resting ulti-
mately on philosophical preference rather than empirical validation. The
origins of the universe, of life, and of humans, it may be argued, are beyond
formal scientific investigation. If this is so, then the way forward may be no
different from the way back – a procession of untestable theories.

However, the link between the origins of human behaviour and other
cosmological problems may perhaps show a means of escaping this rather
despairing scenario (Foley & Dunbar 1989). Physicists and mathematicians
studying the origins of the universe are faced with a very similar problem to
that of palaeoanthropologists – the investigation of events that took place
millions of years ago, occurring under conditions very different from those of
today. These events cannot be directly observed but must be inferred from the
observation of remnant effects. The universe is in a sense a fossil of the big
bang and subsequent developments. Indeed, cosmologists study events even

more remote than human origins, and conditions that differ greatly from those found today. And yet physicists have managed to reconstruct what happened and to reveal the principles underlying events at the beginning of the universe. The approach they have used to achieve this may provide a useful pointer to the way forward for palaeoanthropologists.

The answer seems to lie in the use and nature of theory, and in its articulation with empirical study. Relativity and quantum mechanics – purely theoretical formulations based on mathematical principles – have allowed a fairly narrow set of models to be constructed. These have enabled certain possibilities to be ruled out. They have also permitted precise predictions to be made about the outcomes arising from certain models, and hence the specification of empirically observable entities and relationships that will occur only if the conditions outlined in the models are fulfilled. In other words, moving from simple description and classification to both prediction and explanation is dependent upon the development of theory and the disciplining of that theory to the constraints of empirical investigation.

It is obvious that the complexities of the biological world pose a vastly different set of problems from the certainties of physics. Biologists, especially those working at the 'softer end' of ecology and behaviour, cannot even start to claim that their theories are as powerful as those of the cosmologists. Indeed, it may be argued that biology, dealing as it must with events strongly affected by historical contingency, can never hope to rival the description of the physical world. However, physical scientists have two important lessons for us. The first is that considerable progress can be made not by seeking a single answer in one leap but by eliminating certain possibilities. It may not, at this stage, be possible to determine whether early hominids were monogamous or polygamous, but it may be possible to exclude other alternatives such as polyandry or asociality on theoretical grounds alone (Foley & Lee 1989, Lee 1988/9). The second is the link between theories, models, and empirical observation. Theories expose the principles we think underlie the events we are studying. These may be transformed into models, which are essentially conditional statements that arise out of the theoretical axioms. Their utility is based on the extent to which they can specify outcomes to be expected if certain conditions are fulfilled. If those outcomes have empirical manifestations, then formal testing becomes a practical possibility.

As discussed above, the expansion of evolutionary theory, and especially the development of evolutionary and behavioural ecology, now makes it possible for biologists to construct formal theories – sets of expectations – relating to the behaviour and adaptations of complex organisms. While these are not complete, and may be particularly problematic when applied to modern humans, they nonetheless provide a series of expectations. Only with these expectations can we hope to measure the extent to which humans may or may not conform to the general principles underlying the behaviour of biological organisms. Furthermore, for those studying living organisms, they can be formulated in terms of empirically testable (and quantifiable) hypotheses (Dunbar 1989). The extension of these to events and situations in the past is, of course, a formidable technical problem, but it remains the only option. That this is not a simple attempt to reduce all behaviour to a system of energetics is perhaps evidenced by the way in which behavioural ecology is itself, while not abandoning its basic principles, moving towards incorporating cognitive and other more complex parameters into its framework (Byrne & Whiten 1986). The same should be expected for behavioural palaeoanthropology. Klein gives an example of this (Ch. 5) when he attempts to build a model of the structure of human thought which is consistent with both evolutionary principles and Lévi-Straussian structuralism.

In elaborating these methodological principles two further points should be made. The first is a question of discipline. Major advances in science are seldom made by asking questions in a very general form. Crick and Watson did not solve the problem of the structure of genes by asking metaphysical questions about the nature of heredity but by recognizing that certain key properties must be involved; this in turn led them to focus on the exact molecular structure of chemicals in the nucleus of the cell (Crick 1989). Similarly, we should not be asking what the origins of human behaviour are, but should be looking instead for the precise properties of humans that we seek to understand. These may turn out to be energetic, cognitive, or even thermoregulatory – at present we simply do not know.

Second, the key process is that of inference. The task at hand involves not direct observation of what we are trying to understand but indirect inference from other observations. The key to inference lies in the fact that we are using posited theoretical principles to move from what is observable to what is not. Any study of events in the past must proceed in this fashion. That the advances made in cosmology are greater than those made in palaeoanthropology can be explained by the fact that inference in physics is more certain, as the underlying principles of physics are simpler and more verifiable than those in biology, especially behavioural biology. However, it is very important to recognize that inference as the path to knowledge is not confined to the study of the past. All knowledge is based on inference from observations; this is as true of experimental sciences (for example, inferring chemical structure from X-ray diffraction patterns) and the social sciences (for example, inferring social structure from discussions with informants) as it is of sciences investigating past events. Differences between 'neo' and 'palaeo' sciences derive from the solidity and testability of the links of an inferential chain, not from any inherent differences in the way in which knowledge is constructed.

The way forward, then, lies in developing a judicious mixture of theory-building and empirical observation. Is it possible to be more precise about the nature of these? As already stated, the theoretical framework now available for behavioural palaeoanthropologists is considerably more sophisticated than it was in the past, and has been elaborated in considerable detail elsewhere (Foley 1987a, Kinzey 1987, Mithen 1989). The construction of models from this theoretical framework is also an area that has been strengthened, in particular by the application of computer simulations and other quantitative techniques (Dunbar 1989). Zubrow (Ch. 6) provides an elegant example, showing both how it is possible to incorporate individual and population variations, which must be the key to past events, into models; and how small changes in these variables can have major effects – in this case, the replacements of Neanderthals by anatomically modern humans. These models in turn have enabled us to gain a better idea of what sort of data may be critical to the further investigation of the origins of human behaviour.

One example may be briefly mentioned here to illustrate the interrelationship between theory, model-building and empirical observation, and hence the way in which knowledge about early hominid behaviour may be advanced. Body size has become a central focus of evolutionary ecology. It is now clear that a large number of energetic, physiological, and behavioural parameters greatly affect, or are affected by, body size (Peters 1983, Martin 1983). These include, for primates and many other organisms, brain size, metabolic rate, longevity, reproductive rates, patterns of growth, sexual dimorphism, diet, and patterns of locomotion, as well as many others (Harvey *et al.* 1987). In many cases the nature of these relationships has been quantified.

What is particularly relevant is the fact that body size is a trait that may be inferred from observations of fossils (McHenry 1989). This means that access to a wide range of behavioural and ecological characteristics of an extinct organism is possible. Changes in body size during the course of hominid evolution, or differences in body size across extinct taxa, can therefore be used to infer other attributes that may be of great evolutionary significance – degree of encephalization, energetic requirements and costs, patterns of ranging behaviour, and so on. Apart from body size, similar relationships between empirically observable entities and other evolutionarily significant parameters are currently being investigated – for example, growth and development (Bromage & Dean 1985, Smith 1989), longevity and demography (Trinkaus & Thompson 1987), and species diversity (Foley, in press).

The selection of these predominantly palaeontological examples undoubtedly reflects my own concerns and biological orientation. Other parameters such as technology should be equally to the fore in constructing such methodologies. Indeed, Wynn (Ch. 4) provides an explicit example of how such inferential links can be made using technological evidence. However, it is probably the case that the more biologically based aspects of palaeoanthropology have developed quantitative modelling rather more fruitfully, until now, than has been the case in palaeolithic archaeology. To some extent this reflects the more formally constructed nature of theory in evolutionary ecology compared with archaeology (contrary to the position outlined in Clark 1989). More important, though, it highlights the need to incorporate the material archaeological record, by far the most abundant form of direct evidence of the past, into behavioural palaeoanthropology to a greater extent than has hitherto been the case (but see Wynn, Ch. 4). Palaeolithic archaeology needs to become much more integrated with evolutionary biology.

The methodological framework outlined here assumes that the questions of interest to behavioural palaeoanthropologists are relatively clear-cut.

However, this is far from true. To return to a point made earlier, in order to advance our understanding of the origins and evolution of human behaviour, we need to specify tightly formulated questions. These do derive, without doubt, from a general understanding of what 'being human' means – that is, an intelligent, behaviourally flexible, technologically dependent, highly communicative and conscious species that lives in extensive and complex social contexts in very large numbers. This broad definition sets the parameters for the specific questions to which we need to know the answers:

> What is the pattern of development in human intelligence, and more particularly, what are the types of mental skills that constitute our mind?
> To what extent were hominids other than modern humans flexible in their behaviour? Can any differences be quantified, and do they vary according to different types of behaviour (for example social versus foraging behaviour)?
> What is the pattern of technological development during the course of human evolution, and especially to what extent does technology in prehuman hominids differ in its pattern of generation and use from that in modern humans?
> When did the communicative skills of humans develop, and how is such communication related to the capacity for conscious and self-reflective thought?
> What are the patterns of social organization of hominids? This is partly a question about group size, but more interestingly concerns relationships between and within sexes and across age categories, and the degree to which they varied within and between populations.

While these questions approach a greater degree of specificity, they are still extremely general. The ways in which they may be made even more precise can illustrate some important issues in behavioural palaeoanthropology. For example, underlying each one of the above questions are a number of issues, such as:

> Under what ecological conditions would, for example, certain mental skills be useful?
> Which hominid taxa exhibit particular types of mental skills?
> What were the consequences of their evolution for other attributes of hominid lifestyle?
> How were they developed in an ontological sense?

Foley (Ch. 2), while focusing on analytical problems associated with culture, attempts to sketch out a model that links human behavioural and cognitive evolution to ecological conditions.

It is worth highlighting some aspects of our current understanding of hominid evolution, as they provide useful insights into what may have happened in hominid evolution. First, it is abundantly clear that human evolution is not a ladder-like progression from an ape ancestor to modern humans. Rather, it is a bush of radiating populations and species, each of which may well have had characteristics unique to itself. To force the variation observable in the archaeological and palaeontological record into a linear framework is likely to be extremely misleading. Perhaps the most striking evidence for this lies in the fact that anatomically modern humans may in fact have predated the classic manifestations of the Neanderthals (see Mellars & Stringer 1989). Second, human evolution may not have been a pattern of constant change; key behavioural changes may have been specifically located at particular points in time and space. For many, an example of this would be the origins of modern humans, who may have represented a radical departure from other hominids, rather than a continuation of existing trends (Foley 1989). Third, there may be considerable differences in the way in which similar behaviours were generated in different hominid taxa. For example, the actual process of manufacturing a bifacial handaxe may lie within the technical compass of a variety of species of hominid – and hominoid (McGrew, Ch. 1) – but the way in which each of these taxa executed the technology may have been very different. And finally, hominid morphology shows considerable regional continuity for some periods, with distinctive trajectories of change occurring, and a similar claim may be made for the archaeological record (Foley 1987b). Here it is possible to see novel selective pressures, which may have been the same in different regions, interacting with unique phylogenetic factors to produce divergent patterns of evolution. The implication is that there may be considerable historical contingency in the pattern of hominid evolution. As Habgood shows (Ch. 7), even the morphological evidence requires dextrous integration of historical, genetic, demographic, and ecological factors.

These very brief examples indicate an important trend in the study of the origins and evolution of human behaviour that should lead to future research – that is, the need to be increasingly precise. As mentioned earlier, it is no longer possible to refer generally to hominids or early hominids; behavioural models must be specific to time, place, and taxon. Furthermore, any model must incorporate both proximate (for example, developmental) and ultimate or selective parameters. And finally, any model must tackle the problem of the interaction of immediate functional causation with problems of historical contingency. In this way, behavioural palaeoanthropology may move towards a more precise understanding of the way in which the unique features of the modern human species evolved.

References

Bateson, P. P. G. 1988. The active role of behaviour in evolution. In *Evolutionary processes and metaphors*, M. W. Ho & S. W. Fox (eds), 119–207. New York: John Wiley & Sons.

Bromage, T. G. & M. C. Dean 1985. Re-evaluation of the age of death of immature fossil hominids. *Nature* **317**. 525–7.

Byrne, R. & A. Whiten 1986. *Machiavellian intelligence*. Oxford: Clarendon Press.

Clark, G. A. 1989. Alternative models of Pleistocene biocultural evolution: a response to Foley. *Antiquity* **63**, 153–62.

Crick, F. 1989. *What mad pursuit*. Harmondsworth: Penguin.

Dunbar, R. I. M. 1988. *Primate social systems*. London: Croom-Helm.

Dunbar, R. I. M. 1989. Ecological modelling in an evolutionary context. *Folia primatologica* **53**, 235–46.

Foley, R. A. 1987a. *Another unique species: patterns in human evolutionary ecology*. Harlow: Longman.

Foley, R. A. 1987b. Hominid species and stone tool assemblages: how are they related? *Antiquity* **61**, 380–91.

Foley, R. A. 1989. The ecological conditions of speciation: a comparative approach to the origins of anatomically modern humans. In *The human revolution: behavioural and biological perspectives on the origins of modern humans*, P. A. Mellars & C. B. Stringer (eds), 298–320. Edinburgh: Edinburgh University Press.

Foley, R. A., in press. How many hominid species should there be? *Journal of Human Evolution*.

Foley, R. A. & R. I. M. Dunbar 1989. Beyond the bones of contention. *New Scientist* **124** (1686), 37–41.

Foley, R. A. & P. C. Lee 1989. Finite social space, evolutionary pathways and reconstructing hominid behavior. *Science* **243**, 901–6.

Harvey, P. H., R. D. Martin & T. H. Clutton-Brock 1987. Life histories in comparative perspective. In *Primate societies*, B. B. Smuts et al. (eds), 181–96. Chicago: University of Chicago Press.

Holmes, E. C., G. Presole & C. Saccone 1989. Stochastic models of molecular evolution and the estimation of phylogeny and rates of nucleotide substitution in the hominoid primates. *Journal of Human Evolution* **18**, 775–94.

Humphrey, N. K. 1976. The social function of intellect. In *Growing points in ethology*, P. P. G. Bateson & R. A. Hinde (eds), 303–17. Cambridge: Cambridge University Press.

Kinzey, W. G. (ed.) 1987. *The evolution of human behavior: primate models*. Albany: State University of New York Press.

Lee, P. C. 1988/9. Comparative ethological approaches in modelling hominid behaviour. *Ossa* **14**, 113–26.

Martin, R. D. 1983. Human brain evolution in ecological context. James Arthur Lecture on the Evolution of the Human Brain, American Museum of Natural History, New York.

Medawar, P. B. 1967. *The art of the soluble*. London: Methuen.

Mellars, P. A. & C. B. Stringer (eds) 1989. *The human revolution*. Edinburgh: Edinburgh University Press.

Mithen, S. 1989. Evolutionary theory and post-processual archaeology. *Antiquity* **63**, 483–94.

McGrew, W. C. 1990. Chimpanzee material culture: implications for human evolution. Unpublished PhD thesis, University of Stirling.

McHenry, H. M. 1989. New estimates of body weight in early hominids and their significance to encephalization and megadontia in robust australopithecines. In *Evolutionary history of the robust australopithecines*, F. E. Grine (ed.), 133–48. New York: Aldine.

Peters, R. H. 1983. *The ecological implications of body size*. Cambridge: Cambridge University Press.

Shennan, S. 1989. Cultural transmission and cultural change. In *What's new? A closer look at the process of innovation*, S. E. van der Leeuw & R. Torrence (eds), 330–46. London: Unwin Hyman.

Smith, B. H. 1989. Dental development as a measure of life history in primates. *Evolution* **43**, 683–7.

Standen, V. & R. A. Foley (eds) 1989, *Comparative socioecology: the behavioural ecology of humans and other mammals*. Oxford: Blackwell Scientific.

Trinkaus, E. & D. D. Thompson 1987. Femoral diaphyseal histomorphometric age determinations for the Shanidar 3, 4, 5, and 6 Neanderthals and Neanderthal longevity. *American Journal of Physical Anthropology* **72**, 123–9.

1 Chimpanzee material culture: what are its limits and why?

W. C. McGREW

Introduction

It is a truism to say that behaviour and ideas do not fossilize. Hence, reconstruction of the origins of culture depends on artefacts and other remnants, the use and meaning of which are then inferred. Such inference may seem simple in principle but it is difficult in practice, for several reasons: first, cultural objects are not always distinguishable from natural ones; second, only a subset of enduring objects remains while perishable ones have been lost; and third, deposition and sometimes recovery is nonrandom, and so what remains in the archaeological record is biased. The upshot of this can be summed up in two aphorisms: 'Absence of evidence is not evidence of absence' and 'Presence proves only possibility, not probability'. In the case of the former, one could not infer that early hominids did not use digging sticks, as these tools would inevitably be lost to us. In the case of the latter, concentrations of fragments of fired clay need not imply human agency, as natural sources may be equally likely.

These difficulties may be eased by looking for the closest living approximation to the extinct hominoid forerunner. Such referential models (Tooby & DeVore 1987) will not be identical, obviously, but the closer the fit, the better. The big advantage thus gained is that directly observable behavioural data will be available to supplement the artefacts. At the very least, this shows the minimal capacity of an early hominid. It is necessary, but not sufficient, evidence.

Consider an idealized example. An ape is seen to make and use a stone tool which is indistinguishable from a similar object thought to have been made by an early hominid. This may mean nothing more than limited but certain knowledge of one way in which that artefact could have come about. However, it is a tremendous advance, because one now has available for study the behaviour and mind of the user. For an empiricist, this is worth all the speculation, however fascinating, in the world.

So, how to choose the best model? Even a glance at the palaeoanthropological literature shows no consensus. Some models rely solely on living human beings and exclude other species. Such approaches rightly favour tropical, open-country-living, hunting-and-gathering people, especially in Africa (Musonda, Ch. 3, this volume). Other models follow either homology, and make use of nonhuman primates, or analogy, and make use of social carnivores. My aim here is to focus on the common chimpanzee, *Pan troglodytes*, in both ways.

The chimpanzee is apt for several reasons. It is well studied, both in the wild and in captivity. Anatomically and genetically, it is our closest living evolutionary relation. Most important, it is a culture-bearing creature in its own right. (This last point is debatable and may be a curse as well as a blessing; see Foley, Ch. 2 this volume, McGrew 1990.)

Of course, as others have used the chimpanzee model before, the reader might well ask that more can usefully be said. The answer is that if the most complete and thoughtful earlier work was done by Tanner (1981, 1987, 1988), then new findings have already rendered it out of date. The new knowledge comes from several sources. In captivity, unprecendentedly rich and creative experiments and observations have been done with chimpanzees. In nature, several new field sites and studies have been developed, as well as older ones expanded. In palaeoanthropology, new techniques for analysis of artefacts and fossils yield data never before available and, in some cases, not even imagined. Archaeological evidence is much improved in quality as well as in quantity, especially in terms of careful, systematic collection. Perhaps most important, analysis is tighter and more rigorous, relying on explicit, step-by-step argument and stated, falsifiable hypotheses, rather than seductive but slippery scenarios. Thus my aim, restated, is to answer the following question: given recent knowledge, is the chimpanzee a better or worse model for human evolution than before? In tackling this problem, the catchier question posed in this chapter's title should be dealt with too.

A cautionary note: culture is not a concrete entity but a mental construct. It is a set of concepts and as such cannot truly evolve. However, capacities for culture can evolve in organisms, and manifestations of culture, such as artefacts, can evolve in the sense of showing changes in design, features, etc. Being tangible, material culture is the easiest point at which to start retrospective analysis, but it is not enough to stop at the material. To be used meaningfully, the term culture implies associated symbol-use by the culture-bearer. Thus reconstructing cultural evolution may start with objects for convenience, but must carry on to assess the symbolic significance that is implicit in the artefacts.

On another front, there is disagreement about the various extinct forms for which the chimpanzee has been proposed as a model. At least four have been put forward in print: first, a Miocene stem-form of ancestral ape; second, a Mio-Pliocene ancestral hominoid, which cannot be assigned confidently to either Pongidae or Hominidae; third, the first recognizable Pliocene hominid; and fourth, a later, Plio-Pleistocene hominid, the first indisputable member of the genus *Homo* (McGrew 1989). The first of these may correspond to *Proconsul*; the second is unknown; the third, an australopithecine or *H. habilis*; the last, *H. erectus*.

For reasons elaborated upon elsewhere (McGrew, 1990), it seems likely that only the middle two need be taken seriously. The first is too conservative, in that it was probably a precultural form, more like a present-day gibbon. The last is too advanced, in that living chimpanzees arguably show neither the cultural nor the bodily similarities needed to draw direct comparisons with large-brained *Homo*. Of the middle two, the ancestral hominoid is to be preferred on grounds of caution, but it must be an underestimate if the living chimpanzee is a culture-bearer (unless one assumes cultural devolution). It is hardly likely that pongid culture has stood still over the last 6 million years. On the other hand, if the living chimpanzee most resembles an australopithecine, then we must be faced with cultural convergence, or parallelism, since phylogenetically the pongid and hominid lines had already diverged, or, to be more precise, at least one radiation had occurred.

Culture and symbols

As might be expected, culture is as hard for anthropologists to define operationally (as opposed to theoretically) as intelligence is for psychologists or language for linguists. It is not enough to say that culture includes all things human, for a concept that explains everything explains nothing; it is not heuristic. Thus a working definition is crucial if one is to tackle the evolutionary transition from a precultural to a cultural state. If the change had intermediate, protocultural states, then precision and explicitness are all the more important. For the sake of this chapter, culture will be thought of as being made up of observable actions and inferrable thoughts. The former has been covered in an earlier work, using a set of eight criteria: innovation, dissemination, standardization, durability, diffusion, tradition, nonsubsistence, and natural adaptiveness (McGrew & Tutin 1978). Chimpanzees show all of these, but it is the last which needs attention here.

Cultural organisms must have the mental abilities to create and use symbols; otherwise, one would have pseudoculture, which is essentially mindless, social learning. In other words, self-aware symbol-use is a necessary prerequisite of culture. (Contrary to what is sometimes said – for example, by Washburn and Benedict (1979) – it is symbol-use and not language which is critical. The two are not synonymous, as language is only a subset of symbol-use. This is clear from studies of nonverbal humans such as autistic children.) So, do chimpanzees show self-awareness and use symbols?

The answer on both points seems to be yes. Gallup's (1970) elegant experiment on chimpanzees recognizing themselves in a mirror has been much repeated and elaborated upon. Woodruff and Premack (1979) have shown that chimpanzees seek to deceive humans both by omission and by commission. Savage-Rumbaugh *et al.*'s (1978, 1980) series of careful studies has shown chimpanzees using simple symbols to label, sort, and ask for objects such as food or tools. They do so among themselves, in the absence of humans. All of these demonstrations come from laboratories and so remain to be confirmed in nature, but the capacities are clearly present and are used in social life (de Waal 1982).

Environment and adaptation

Despite evidence to the contrary, prehistorians continue to think dichoto-
mously about the correlation between environment and hominoids. Living
chimpanzees and their pongid ancestors are classed as forest-dwellers, while
hominization is linked with adaptation to the savanna. Given this neat
division, chimpanzees would seem to be precluded as evolutionary models for
any hominid on basic ecological grounds. In fact, recent findings show that
neither distinction holds. While the picture of an African Plio–Pleistocene
vegetational mosaic of scrub, deciduous open woodland, grassland, and scant
gallery forest remains typical, this was punctuated by periods of expansion of
humid rainforest (Williamson 1985).
     More to the point, wild chimpanzees live in hot, dry, and open environ-
ments today, both in East and West Africa. Long-term studies have been
carried out at Mount Assirik, Senegal, where less than 3 per cent of the surface
area is forested (McGrew et al. 1981). Climatologically, hydrologically, flori-
stically, and faunistically, this is a savanna. At the least, the studies show that
chimpanzees as a species are much more adaptable than usually credited, and so
must be eligible as possible models for either forest-living ancestral hominoids
or savanna-living hominids.
     More pertinent to the origins of culture is the extent to which the
chimpanzee toolkit (and hence by analogy the ancestral form's material
culture) varies with environment. While some aspects of chimpanzee material
life such as nest-building are largely constant (see p. 19), variation in other
aspects occurs across populations. Tools used to obtain termites for food differ
over three widespread sites (McGrew et al. 1979). Moreover, such variation
also occurs between communities in the same population (McGrew & Collins
1985). In both cases, some differences reflect contrasting features of habitat (for
example, availability of prey), while others appear to reflect contrasting social
customs (for example, preferences for raw materials). Finally, pan–African
comparison of techniques by which a specific food item, the oil palm nut
(*Elaeis guineensis*), is processed and eaten by chimpanzees, shows a range of
cultural complexity (McGrew 1985). Some populations ignore the nuts,
others eat only the outer energy-rich husk, and other also use stone tools to
extract the protein-rich kernel.
     All of this sounds familiar when one recalls early hominid cultural tradi-
tions described on the basis of different lithic industries (Leakey 1975). Recent
reassessments (Toth 1985b) are more cautious in their interpretations and
inferences and are even closer to the chimpanzee model. For example, early
hominids at Koobi Fora may not have depended on stone tools and may have
used them only in certain habitats. In some cases, prehistorians and primatolo-
gists working independently, with different data, have come to very similar
conclusions: for example, that design of tools is demonstrably a function of the
size, shape, and mechanical properties of the raw materials (Jones 1981,
McBeath & McGrew 1982, Boesch & Boesch 1983).
     In summary, recent ecological studies of wild chimpanzees and their

material culture strengthen rather than weaken the case for their being used as models of human cultural evolution. Also, they underline the importance of environmental variables as influences on culture. However, they do not yet allow palaeoanthropologists to match the chimpanzee model with a particular ancestral form.

Diet

Overall, diets of wild chimpanzees and ancestral hominoids and hominids look more and more similar. Both are omnivores, or, more exactly, mainly frugivores with a taste for opportunistic faunivory.

All closely studied populations of chimpanzees are known to eat animal as well as plant matter. Short- and long-term studies of both tamed and untamed wild apes in a variety of habitats show them to eat social insects and small-sized mammals such as monkeys and young ungulates (Teleki 1973, McGrew 1983, Takahata et al. 1983, Boesch & Boesch 1989). Also, cannibalism, once considered aberrant, is now recognized and explicable (Goodall 1977, Nishida & Kawanaka 1985). Thus, meat-eating is species-typical.

However, there are dietary differences between ape populations in the species of prey chosen, and these seem to be understandable only in terms of cultural differences (McGrew 1983). Techniques for getting meat vary too. Chimpanzees stalk as well as stumble upon prey, and after seizing the victim may kill it in several ways (Teleki 1973, Boesch & Boesch 1989). Sometimes extractive foraging of hidden prey occurs: for example, chimpanzees at Mount Assirik winkle out bushbabies (*Galago senegalensis*) from their sleeping holes. If in the company of other apes, intense scrounging and sharing usually follows. Even eating shows special features: bites of meat are eaten with mouthfuls of leaves, a habit not yet understood. Chimpanzees pirate prey from other predators: for example, freshly filled bush-buck fawns (*Tragelaphus scriptus*) are taken from baboons (*Papio anubis*) (Morris and Goodall 1977). Significantly, true scavenging also occurs, when the carcass of an ungulate not seen or heard to be killed is eaten when found later (Hasegawa et al. 1983).

Cross-population differences also exist for plant foods eaten (Nishida et al. 1983), but fruits remain the staple of chimpanzee diet, from the wettest to the driest habitats. (McGrew et al. 1988, Nishida & Uehara 1983). Every place, however dry, at which chimpanzees have been studied for a long time (that is, over several annual cycles) has yielded fleshy fruits in the gallery forest, at least at some times of year (Baldwin 1979).

What wild chimpanzees do *not* eat is equally important, especially as negative evidence is now strong from a few very long studies. For animal foods they avoid fast-moving, solitary prey, which are, in effect, more trouble than they are worth, such as nonsolitary insects, reptiles, amphibians, and most small mammals. Moreover, mammals weighing more than about 15 kg are not preyed upon. This is true even of species whose young are taken, such as bushpigs (*Potamochoerus porcus*). This pointed omission is in all likelihood due to the adults' abilities to avoid or repel predatory apes, which kill only by grabbing their prey.

For plants, what is ignored is even more telling: for example, underground storage organs such as tubers and rhizomes are not eaten, even when present and exploited by sympatric primates such as *Papio papio* at Mount Assirik. This omission is notable given the proposed prominence of roots in hominid adaptation to the savanna (Hatley & Kappelman 1980). Similarly, chimpanzees feed only rarely on the seeds of grasses. This is true even on savannas where edible grasses abound and are heavily exploited by baboons (McGrew *et al.* 1988, Sharman 1981). It seems likely in both cases that the frugivorous ape is outdone by more dentally specialized competitors.

Recent palaeoanthropological evidence of diet is impressive but inconclusive. Happily, much data has replaced earlier speculation. Butchery and presumed carnivory by tool-using Plio-Pleistocene hominids is now known from cut marks on the bones of large herbivores. At least some of these have been found close by hominid fossils or artefacts (Shipman & Rose 1983, Potts 1984b). However, other data, such as patterns of damage to bones in marrow-extraction, are equivocal (Bunn 1981). As yet little can be said about feeding on invertebrate prey at any time, or about feeding on vertebrates before stone tools emerged in hominid evolution. In principle, microwear on the teeth of the hominoid could yield signs of, for example, bone-crushing, or scanning electron microscopy of the bones of vertebrates might yield distinctively human tooth marks, but these experiments remain to be done, or are inconclusive.

Evidence of nonfossilizing plant foods in the diet is even more tenuous. Thus it is not surprising that results and conclusions conflict. On the basis of microscopic tooth-wear, Walker (1981) concluded that robust australopithecines were fruit-eaters, and not grass-, leaf-, or bone-eaters. On the basis of dental anatomy and biomechanical analysis of chewing, Lucas & Corlett (1985) concluded that the same creature was a specialized eater of grass seeds, legumes, and roots. Thus the chimpanzee seems to be either the most or the least apt model for this early hominid! More detailed comparative data are available on a Miocene hominoid, *Sivapithecus indicus*. According to Teaford & Walker (1984), its pattern of dental microwear is indistinguishable from that of *Pan*, but differs from other living primates which are more specialized for hard-object- and leaf-eating. In contrast, on the basis of their thick molar enamel, Kay (1981) decided that the ramapithecines (including *S. indicus*) ate hard fruits, seeds, and nuts. There is agreement that Miocene hominoids were not grass-eaters (see also Covert & Kay 1981).

Overall, the patchy evidence now available suggests that the diet of chimpanzees may resemble that of an ancestral hominoid more than an intermediate hominid. Only further systematic and experimental studies (Peters 1982) will clarify the picture. More data are needed, not more speculation. Whatever the uncertainty about the fossil evidence, though, the chimpanzee looks markedly better than any other living primate as a dietary analogue.

Home bases

Early field studies of chimpanzees stressed their nomadism and lack of fixed or even revisited living or sleeping sites. This is easy to understand in woodland and forest, where food and trees are many and scattered. More recent studies of these apes in open, high-seasonal habitats show a different picture. At Mount Assirik, by the end of the seven-month dry season, chimpanzees were sleeping only in the narrow strips of gallery forest (Baldwin *et al.* 1982). Moreover, within this forest, sleeping sites were concentrated around the few remaining sources of clean water.

Similarly, early reports concluded that sleeping platforms ('nests') built by apes were largely stereotypes in their making and resulting form. However, if one compares nests point by point across populations, differences emerge, even in such seemingly arbitrary features as the proportion which were open to the sky instead of sheltered by overhanging foliage (Baldwin *et al.* 1981).

Finally, chimpanzees in their daily activities do not wander constantly or aimlessly. Travel (as opposed to feeding on the move) follows paths much of the time, and favoured resting spots are used again and again. At some times of year, certain resources, such as termite mounds or hammers and anvils, may be visited several times a week (Boesch & Boesch 1984, McGrew and Collins 1985).

Palaeoanthropologists seeking to interpret evidence of home bases in the archaeological record focused on safety and food-sharing as key reasons for base camps (Isaac 1978). More recent reinterpretations of such concentrations of bones and tools argue for caution, as other agents such as flowing water, natural clusters of dead animals, or scavenging carnivores may have been responsible (Potts 1984a). Even accumulations of stone tools may have been occasionally used as caches rather than longer-term occupation sites. Thus there is no reason to assume that either early *Homo* or earlier ancestral hominoids were more home-based than living chimpanzees are.

Technology

Chimpanzees are skilful makers and users of tools. They make a variety of tools from a variety of raw materials to serve a variety of purposes. Different communities have different toolkits, both within and across populations. The basic picture is well known (Goodall 1964), but recent findings refine it. For example, diffusion of a tool-use pattern has been seen: for example, termite-fishing females have migrated from one group to another in the Mahale Mountains, Tanzania (Takahata 1982). Telling cases of tool-use in hunting have been seen, such as an adult male throwing a rock to break up a stand of bushpigs, allowing piglets to be grabbed and eaten (Plooij 1978). Spontaneous appearance and rapid dissemination of hammer-stone-use has been followed in a group of 16 chimpanzees (Hannah & McGrew 1987). Chimpanzees in the Tai Forest, Ivory Coast, use hammer-stones of differing raw materials and dimensions to crack open nuts of different species, and show sex differences in doing so (Boesch & Boesch 1983, 1984). Finally, chimpanzees will transport raw materials, tools, and items to be processed for hundreds of metres before use, even if the resource or place of use is out of sight (Boesch & Boesch 1985, Hannah & McGrew 1987). Far from being exhausted, the technological capacities of chimpanzees continue to be revealed.

However, there are certain things that the apes have not yet been seen to do. In most cases this reflects the limits of their naturally endowed, mostly dental, features. They do not make flaked stone tools, presumably because they already possess cutting edges on their canine teeth. But they do use hammer-stones to open nuts, presumably because their molar teeth are not robust enough to break the hardest-shelled species without risk of damage. They have enough strength to dismember small prey such as monkeys by hand, without butchering tools, but resort to flimsy probes when the prey is an underground termite which requires delicate extraction. There are 'gaps' too, such as the lack of digging sticks to get roots, as noted above. Neither do they make or use shelters or containers in nature, though in captivity they readily take to both. They do not use missiles or ladders to bring down or gain access to, for example, out-of-reach fruit in the wild, but will do so if taught by humans.

Attributing function to the finished product, the artefact in the archaeo-logical record, is notoriously difficult. Reconstructing the making of such a tool is even more difficult, as this so far has depended on experimental replication by knappers (Jones 1981). Occasionally, ingenious approaches may yield new knowledge from old items: for example, microwear on tool edges can be an indicator of what material was worked (Keeley 1977); the sequence of flake removal can reveal whether the worker was right-handed or left-handed (Toth 1985a). However, such studies have so far tended to concentrate on artefacts from later in cultural evolution, typically those of *H. erectus*, rather than on those from earlier forms.

Finally, the most challenging type of inference is that of the mind behind the act which produces the artefact. This twice-removed operation is fraught with uncertainty. Gowlett's (1984) 'procedural templates' (though really flow charts of action rather than thought) at least make explicit the possible sequences of manufacture from start to finish.

Perhaps the most ambitious attempt to tackle the problem of extinct intelligence is that of Wynn (1979, 1981, Ch. 4, this volume). He uses the genetic epistemology of Jean Piaget, the Swiss polymath, to re-create the minds of the makers of tools from the tools' topological attributes. For Oldowan tools he concludes that no more intelligence than that of a living chimpanzee would be needed. For Acheulean tools he goes to the other extreme and claims that their making required a level of intellect no less than that of adult *H. sapiens*. However, a closer look at the four mental operations posited – whole–part relations, qualitative displacement, spatiotemporal substi-tution, and bilateral symmetry – shows all of them to be involved in the chimpanzees' making of probes for termite-fishing (McGrew 1990).

Conclusions

New knowledge from the last decade makes the chimpanzees a better model of the origins of human culture than ever before. To be more exact, the relative number of points of similarity has increased and those of dissimilarity have declined. For example, specificity of raw materials for tools is independently and unexpectedly confirmed in both ape and ancestor, hence a similarity exists. Scavenging of carcasses of unknown origin has now been seen in chimpanzees, hence an hypothesized dissimilarity has disappeared. Also the relative degrees of similarity have more often edged closer together than moved apart. Diffusion of a tool-use skill, long known in captivity, has now been confirmed in nature. Put another way, more and more hypothetical differences between a living pongid and an extinct form ancestral to the living hominid are now seen to be quantitative rather than qualitative.

Ironically, however, the new knowledge carries with it complications. If chimpanzees are culture-bearers, then they too have a heritage of cultural evolution. If this goes all the way back to a common ancestral hominoid, then African ape and hominid cultural evolution may have gone on in parallel, or even intertwined, for millions of years. Oldowan tools could have been made by apes, not humans (Wynn & McGrew 1989).

More likely, it means analogy is just as important as homology in comparisons between chimpanzees and possible extinct counterparts. Apparent limits on chimpanzee achievements may be cultural, not organic. As such, they are not fixed. If a previously isolated human society which has no written form of language is found, the assumption is made that this is a cultural absence, not an organic one. Now, if one finds a chimpanzee population without stone tool-use, the same sort of interpretation must follow. This means that use of a chimpanzee model for help in reconstructing human evolution can no longer be species-typical. Instead attention must now be paid to ethnographic details in another species. Having learned more about chimpanzee nature, we now know less about its limits.

References

Baldwin, P. J. 1979. The natural history of the chimpanzee (*Pan troglodytes verus*) at Mt Assirik, Senegal. PhD thesis, University of Sterling.
Baldwin, P. J., W. C. McGrew & C. E. G. Tutin 1982. Wide-ranging chimpanzees at Mt Assirik, Senegal. *International Journal of Primatology* 3, 367–85.
Baldwin, P. J., J. Sabater Pi, W. C. McGrew & C. E. G. Tutin 1981. Comparisons of nests made by different populations of chimpanzees (*Pan troglodytes*). *Primates* 22, 474–86.
Boesch, C. & H. Boesch 1983. Optimization of nut-cracking with natural hammers by wild chimpanzees. *Behaviour* 83, 265–86.
Boesch, C. & H. Boesch 1984. Possible causes of sex differences in the use of natural hammers by wild chimpanzees. *Journal of Human Evolution* 13, 415–40.

Boesch, C. & H. Boesch 1989. Hunting behavior of wild chimpanzees in the Tai National Park. *American Journal of Physical Anthropology* **78**, 547–73.

Bunn, H. T. 1981. Archaeological evidence for meat-eating by Plio-Pleistocene hominids from Koobi Fora and Olduvai Gorge. *Nature* **291**, 574–7.

Covert, H. H. & R. F. Kay 1981. Dental microwear and diet: implications for determining the feeding behaviors of extinct primates, with a comment on the dietary pattern of *Sivapithecus*. *American Journal of Physical Anthropology* **55**, 331–6.

de Waal, F. 1982. *Chimpanzee politics*. London: Jonathan Cape.

Foley, R. A. 1991. How useful is the culture concept in early hominid studies? In *The origins of human behaviour*, R. A. Foley (ed.), Ch. 2. London: Unwin Hyman.

Gallup, G. G. 1970. Chimpanzees: self-recognition? *Science* **167**, 86–7.

Goodall, J. 1964. Tool-using and aimed throwing in a community of free-living chimpanzees. *Nature* **201**, 1264–6.

Goodall, J. 1977. Infant killing and cannibalism in free-living chimpanzees. *Folia primatologica* **28**, 259–82.

Gowlett, J. A. J. 1984. Mental abilities of early man: a look at some hard evidence. In *Hominid evolution and community ecology*, R. Foley (ed.), 167–92. London: Academic Press.

Hannah, A. C. & W. C. McGrew 1987. Chimpanzees using stones to crack open oil palm nuts in Liberia. *Primates* **28**, 31–46.

Hasegawa, T., M. Hiraiwa, T. Nishida & H. Takasaki 1983. New evidence on scavenging behaviour in wild chimpanzees. *Current Anthropology* **24**, 231–2.

Hatley, T. & J. Kappelman 1980. Bears, pigs, and Plio-Pleistocene hominids: a case for the exploitation of belowground food resources. *Human Ecology* **8**, 371–87.

Isaac, G. Li. 1978. The food-sharing behavior of protohuman hominids. *Scientific American* **238** (4), 90–108.

Jones, P. R. 1981. Experimental implement manufacture and use: a case study from Olduvai Gorge, Tanzania. *Philosophical Transactions of the Royal Society, London* **B292**, 189–95.

Kay, R. F. 1981. The nut-crackers – a new theory of the adaptations of the Ramapithecinae. *American Journal of Physical Anthropology* **55**, 141–51.

Keeley, L. H. 1977. The function of Palaeolithic stone tools. *Scientific American* **237**, 108–26.

Leakey, M. D. 1975. Cultural patterns in the Olduvai sequence. In *After the australopithecines*, K. W. Butzer & G. Li Isaac (eds), 477–93. The Hague: Mouton.

Lucas, P. W. & R. T. Corlett 1985. Plio-Pleistocene hominid diets: an approach combining masticatory and ecological analysis. *Journal of Human Evolution* **14**, 187–202.

McBeath, N. M. & W. C. McGrew 1982. Tools used by wild chimpanzees to obtain termites at Mt Assirik, Senegal. *Journal of Human Evolution* **11**, 65–72.

McGrew, W. C. 1983. Animal foods in the diets of wild chimpanzees (*Pan troglodytes*): why cross-cultural variation? *Journal of Ethology* **1**, 46–61.

McGrew, W. C. 1985. The chimpanzee and the oil palm: patterns of culture. *Social Biology and Human Affairs* **50**, 7–23.

McGrew, W. C. 1989. Why is ape tool use so confusing? In *Comparative Socioecology*, V. Standen & R. A. Foley (eds.), 457–72. Oxford: Blackwell Scientific.

McGrew, W. C. 1990. Chimpanzee material culture: implications for human evolution. Unpublished PhD thesis, University of Stirling.

McGrew, W. C. & D. A. Collins 1985. Tool-use by wild chimpanzees (*Pan troglodytes*) to obtain termites (*Macrotermes herus*) in the Mahale Mountains, Tanzania. *American Journal of Primatology* **9**, 47–62.

McGrew, W. C. & C. E. G. Tutin 1978. Evidence for a social custom in wild chimpanzees? *Man* **13**, 234–51.

McGrew, W. C., P. J. Baldwin & C. E. G. Tutin 1981. Chimpanzees in a hot, dry, and open habitat: Mt Assirik, Senegal, West Africa. *Journal of Human Evolution* **10**, 227–44.

McGrew, W. C., P. J. Baldwin & C. E. G. Tutin 1988. Diet of wild chimpanzees (*Pan troglodytes verus*) at Mt Assirik, Senegal: I. Composition. *American Journal of Primatology* **16**, 213–26.

McGrew, W. C., C. E. G. Tutin & P. J. Baldwin 1979. Chimpanzees, tools, and termites: cross-cultural comparisons of Senegal, Tanzania, and Rio Muni. *Man* **14**, 185–214.

Morris, K. & J. Goodall 1977. Competition for meat between chimpanzees and baboons of the Gombe National Park. *Folia primatologica* **28**, 109–21.

Musonda, F. B. 1991. The significance of modern hunter-gatherers in the study of early hominid behaviour. In *The origins of human behaviour*, R. A. Foley (ed.), Ch. 3. London: Unwin Hyman.

Nishida, T. & K. Kawanaka 1985. Within-group cannibalism by adult male chimpanzees. *Primates* **26**, 274–84.

Nishida, T. & S. Uehara 1983. Natural diet of chimpanzees (*Pan troglodytes schweinfurthii*): long-term record from the Mahale Mountains, Tanzania. *African Study Monographs* **3**, 109–30.

Nishida, T., R. W. Wrangham, J. Goodall & S. Uehara 1983. Local differences in plant-feeding habits of chimpanzees between the Mahale Mountains and Gombe National Park, Tanzania. *Journal of Human Evolution* **12**, 467–80.

Peters, C. R. 1982. Electron-optical microscopic study of incipient dental microdamage from experimental seed and bone crushing. *American Journal of Physical Anthropology* **57**, 283–301.

Plooij, F. X. 1978. Tool-use during chimpanzees' bushpig hunt. *Carnivore* **1** (2), 103–6.

Potts, R. 1984a. Home bases and early hominids. *American Scientist* **72**, 338–47.

Potts, R. 1984b. Hominid hunters? Problems of identifying the earliest hunter-gatherers. In *Hominid evolution and community ecology*, R. Foley (ed.), 129–66. London: Academic Press.

Savage-Rumbaugh, E. S., D. M. Rumbaugh & S. Boysen 1978. Symbolic communication between two chimpanzees (*Pan troglodytes*). *Science* **201**, 641–4.

Savage-Rumbaugh, E. S., D. M. Rumbaugh, S. T. Smith & J. Lawson 1980. Reference: the linguistic essential. *Science* **210**, 922–5.

Sharman, M. J. 1981. Feeding, ranging and social organisation of the Guinea baboon. PhD thesis, University of St Andrews.

Shipman, P. & J. Rose 1983. Early hominid hunting, butchering, and carcass-processing behaviors: approaches to the fossil record. *Journal of Anthropological Archaeology* **2**, 57–98.

Takahata, Y. 1982. Termite-fishing observed in the M Group chimpanzees. *Mahale Mountains Chimpanzee Research Project Ecological Report*, No. 18.

Takahata, Y., T. Hasegawa & T. Nishida 1983. Chimpanzee predation in the Mahale Mountains, from August 1979 to May 1982. *International Journal of Primatology* **5**, 213–33.

Tanner, N. M. 1981. *On becoming human*. Cambridge: Cambridge University Press.

Tanner, N. M. 1987. The chimpanzee model revisited and the gathering hypothesis. In *The evolution of human behavior: primate models*, W. G. Kinzey (ed.), 3–27. Albany: State University of New York Press.

Tanner, N. M. 1988. Becoming human, our links with our past. In *What is an animal?*, T. Ingold (ed.), 127–40. London: Unwin Hyman.

Teaford, M. F. & A. Walker 1984. Quantitative differences in dental microwear between primate species with different diets and a comment on the presumed diet of *Sivapithecus*. *American Journal of Physical Anthropology* **64**, 191–200.

Teleki, G. 1973. The omnivorous chimpanzee. *Scientific American* **228** (1), 33–42.

Tooby, J. & I. DeVore 1987. The reconstruction of hominid behavioral evolution through strategic modeling. In *The evolution of human behavior: primate models*, W. G. Kinzey (ed.), 183–237. Albany: State University of New York Press.

Toth, N. 1985a. Archaeological evidence for preferential right-handedness in the Lower and Middle Pleistocene, and its possible implications. *Journal of Human Evolution* **14**.

Toth, N. 1985b. The Oldowan reassessed: a close look at early stone artifacts. *Journal of Archaeological Science* **12**, 101–20.

Walker, A. 1981. Diet and teeth–dietary hypotheses and human evolution. *Philosophical Transactions of the Royal Society* **B292**, 57–64.

Washburn, S. L. & B. Benedict 1979. Non-human primate culture. *Man* **14**, 163–4.

Williamson, P. G. 1985. Evidence for an early Plio-Pleistocene rainforest expansion in East Africa. *Nature* **315**, 487–9.

Woodruff, G. & D. Premack 1979. Intentional communication in the chimpanzee: the development of deception. *Cognition* **7**, 333–62.

Wynn, T. 1979. The intelligence of later Acheulean hominids. *Man* **14**, 371–91.

Wynn, T. 1981. The intelligence of Oldowan hominids. *Journal of Human Evolution* **10**, 529–41.

Wynn, T. & W. C. McGrew 1989. An ape's view of the Oldowan. *Man* **24**, 383–98.

2 How useful is the culture concept in early hominid studies?

R. A. FOLEY

Introduction

Culture is a central concept in anthropology. An understanding of the mechanisms of processes of cultural formation, cohesion, maintenance, and change forms a central focus of anthropological studies. The notion of culture has been extensively developed as a unique area of study within the discipline, distinguishing much of anthropology from other branches of the social and behavioural sciences. Furthermore, the concept has acquired a connotation of what is uniquely human – that which distinguishes humans from other animals.

In this context, the culture concept has accumulated a significance in palaeoanthropological studies and, in particular, in models of the evolution of human behaviour. For example, many attempts have been made to identify the origins of truly cultural behaviour in the human evolutionary record, and various markers of these origins have been suggested: meat-eating, cooperative behaviour, food-sharing, home bases, language, symbolic thought, tool-making. The occurrence of evidence for these in the fossil or archaeological record has been used to argue that at this stage – usually placed either in the Early or Middle Pleistocene, and associated variously with the appearance of the genus *Homo* or the species *Homo erectus* – hominids had acquired a new behavioural grade: culture-bearing animals. The existence of this new grade has been employed to suggest that non-Darwinian models of evolutionary change are more appropriate to studies of human evolution: for example, the coevolutionary models of genes and culture developed by Lumsden & Wilson (1981) and Cavalli Sforza & Feldman (1981), among others. In particular, it is often thought that the role of natural selection becomes minimal once culture is established, or that its presence will prevent the operation of certain evolutionary processes (for example, speciation), giving human evolution a distinctive pattern of unilineal and rapid change (Wolpoff 1971).

This chapter lies at one end of a spectrum of views about the value and meaning of the concept of culture. At this extreme is the reductionist approach of evolutionary ecology, which attempts to accommodate new behavioural phenomena within an existing explanatory framework, without recourse to novel entities and processes. At the other is the framework derived from social anthropology, which accepts much more readily hierarchical and emergent entities in developing analytical procedures. The critical question here is which view will be most useful for developing an understanding of the patterns and processes of human behavioural evolution.

Evolutionary ecology attempts to explain phenomena in terms of neo-Darwinian principles of natural selection, principles that place emphasis on individuals rather than larger-scale units, such as societies or cultures. The evolution of culture should therefore be explicable in terms of the advantages it brings to the individual in increased relative reproductive success. In this context the concept of culture is of little analytical value. This chapter attempts to justify this contention.

Definitions of culture

Definitions of culture are almost as numerous as are anthropologists. Kroeber & Kluckholm (1952) have shown the vast range of definitions and usage that exist, and these have probably increased still further in the intervening 38 years. Inherent in a common-sense understanding of culture are such characteristics as learning, nongenetic transmission of information between and among generations, high levels of intra- and especially interpopulation behavioural variability, tool-use and manufacture, and the use of symbolic systems of communication. Beyond this there is some confusion as to whether culture is these observable phenomena or whether it lies in the structure of the mind that makes cultural activities possible. As the purpose of this chapter is to question the utility of the culture concept, and to suggest that the complexity of human behaviour is comprehensible without it, it is not appropriate to develop a new definition. However, the definition Lumsden & Wilson give (1981, p. 3) may be quoted for illustrative purposes: 'culture ... [is] the sum total of mental constructs and behaviours, including the construction and employment of artefacts, transmitted from one generation to the next by social learning'.

The use of culture in palaeoanthropology

Two examples of the use of the culture concept in palaeoanthropology are described here. These should be taken simply as illustrations of the ways in which culture has been employed both in explanatory terms and through modelling terms in human evolutionary studies. The first is Wolpoff's (1971) use of the competitive exclusion principle to argue for a single-species model of human evolution. Wolpoff held that culture was the means by which humans (and hominids) adapted to their environment, and therefore that their niche was defined by the ecological space filled by culture. As culture permits hominids to fill virtually all available ecological space, it could therefore be argued that no two culture-bearing hominid species could exist since these would overlap in their requirements, and so, according to the Gauss competitive exclusion principle, one would come to outcompete the other. Wolpoff's model uses culture, defined as 'structured learning' (although it is difficult to see how learning could not be structured!), in an adaptive context to argue that certain evolutionary options are removed once culture exists – that is, speciation. In his model, culture is used in a dichotomous presence or absence manner, and its presence inferred from attributes that might correlate with the increased capacity for structured learned behaviour: reduced canine dimensions, tool-making, and delayed maturation. In this example culture is used to explain the pattern, supposedly unilinear, of human evolution. Indeed, the model is analytically very powerful, allowing predictions about the nature of the fossil record on the basis of explicit ecological and evolutionary theory.

The second way in which culture has been used in palaeoanthropological studies is for the development of coevolutionary models. These attempt to establish a coevolutionary relationship between genes, evolving through natural selection, and culture, evolving in parallel through some alternative mechanism (Lumsden & Wilson 1981, Cavalli Sforza & Feldman 1981). Essentially these models show that with the establishment of culture, however defined, evolutionary patterns will change in such a way that Darwinian selection can no longer account for patterns or modes of change. These changes can be seen as the actual changes occurring in behaviour (culture), which do not refer back to the genome, and also as the impact of these cultural changes on genetic evolution – hence the coevolutionary model, the interaction between cultural evolution and Darwinian biological evolution. The critical point in these models is that they all start with the assumption that culture constitutes a single entity, often equivalent to the category of 'biology'. Given this assumption, then, twin evolutionary processes can occur, with varying levels of interaction between the biology and the culture. What is problematic about this approach is that it asserts that the development of culture is the creation of a new entity in the evolutionary process. Rather than consisting of a series of epiphenomenal components that have a major impact on the evolutionary process, involving a massive increase in the complexity and variability of the selective environment operating on an individual, culture here is an emergent property and a new evolutionary process.

The inadequacy of culture

Is culture, though, a particularly useful concept in the study of human origins? This is a question relating to the practical utility of culture in analysing a particular evolutionary event. Culture is a composite term, bringing together a whole series of attributes that are important in the way in which humans live today. However, in studying the origins of these attributes, it may not be particularly useful to link them together. We do not know – indeed, this is the very thing we are trying to find out – when any of them first occurred within the hominid lineage. Each of them – the enhanced capacity for learning, speech, tool-making, etc. – may have evolved separately, subject to independent selective forces. Thus to treat them collectively as 'culture' is to remove the possibility that hominids may, in the past, have possessed only part of their present behavioural repertoire, or that repertoire combined in different ways. There is little advantage in using a term that bestows the advantages of a descriptive shorthand (which the term culture certainly does) if it begs the very question we are asking, thus buying descriptive ease at the expense of analytical precision or evolutionary process.

When looked at separately, many of the features that collectively constitute human culture can be found in at least rudimentary form in nonhuman animals. Chimpanzees are known to be tool-users and tool-makers (Goodall 1970, Boesch & Boesch 1983, McGrew, Ch. 1, this volume), and tool-making, as well as the extensive modification of the materials found in the environment (such as occurs in nest-building), are found in other members of the animal kingdom as well as chimpanzees. Although controversial, experiments with chimpanzees have shown them to be capable of systematic use of gestural language (such as American Sign Language), in ways that suggest a grasp of symbols and grammatical structure (Gardner & Gardner 1969, Savage-Rumbaugh et al. 1983). Furthermore, in their natural state, primates employ a wide variety of communicatory systems. Cheney & Seyfarth (1980), for example, have shown that vervet monkeys (Cercopithecus aethiops) use vocalizations in precise ways that are close to what we understand as 'words' – that is, specific sounds that have particular meanings. Learned behaviour is also, of course, extremely widespread among animals, from blue tits learning to open milk bottles (Hinde & Fisher 1951) to birds learning songs from their neighbours (Thorpe 1961) to Japanese macaques acquiring the skill of cleaning the sand off their food (Itani 1958). In each of these instances not only is learning involved, but there is also rapid transmission of information and the development of 'traditions' within populations (McGrew & Tutin 1978).

When treated independently, therefore, most of the features that go to make up the composite world of culture occur elsewhere in the animal kingdom. Use of the term culture in palaeoanthropology obscures this continuity. That they can occur independently, or are combined in ways different from that found in modern humans, or are developed to different degrees, suggests that a reductionist approach is more productive in investigating the origins of modern human behaviour, employing minimalist categories of behaviour. In palaeoanthropology the culture concept makes too many unwarranted and tautological assumptions.

This is not to say that once such complex characteristics as tool-making and communicatory skills have developed they do not result in novel patterns of evolution. It is important, however, to distinguish between causes and consequences in behavioural evolution. As consequences, not causes, of behavioural evolution, they would undoubtedly alter the nature of the selective pressures operating on hominids but not the mechanisms by which selection acts. To assume cultural evolution or coevolution at the outset of an analysis of human evolution is to predetermine that natural selection is inadequate. It is the purpose of an evolutionary ecological analysis to determine whether this is the case, not to prejudge the issue. Behavioural, not cultural, evolution is an adequate term, making the fewest assumptions and providing the greatest flexibility. This does not remove from consideration the characteristics that make humans unique – a vast capacity for learning, innovation and imitation, complex communication, and extreme plasticity of behaviour – but deals with them in ways that make comparisons between species feasible.

Those damned chimpanzees

Culture, then, has three basic flaws as a concept in palaeoanthropology: first, it is too high a level of abstraction to be of much empirical value; second, it assumes a permanent interrelationship of the components that constitute this high level of abstraction, an assumption that is invalid in evolutionary terms; and third, as the basic intention is to define something that is uniquely human, it is constantly redefined in the context of studies of nonhuman primates that show continuities between humans and nonhumans. In the light of these flaws, palaeoanthropologists have the choice of either abandoning the term altogether or else incorporating chimpanzees, and possibly other primates as well, within the realm of 'culture-bearing animals' (see Bonner 1980).

McGrew & Tutin (1978) have opted for the second of these strategies. They have criticized the way in which culture has been defined so as to be virtually synonymous with 'being human', and have instead attempted to set up empirical criteria for defining the existence of culture that allow the behaviour of other animals to be tested for the extent to which it fulfils those criteria. McGrew & Tutin derived their criteria from a challenge put out by the most culture-bound of social anthropologists, Kroeber (1928), as to what he would accept as cultural behaviour in an ape. These criteria are:

innovation	diffusion
dissemination	tradition
standardization	nonsubsistence
durability	natural adaptiveness

According to these criteria, chimpanzees do display cultural behaviour, and would have to be considered as culture-bearing animals. Other primates might also fall within this category. Chimpanzees, therefore, can be said to have demonstrated yet further the continuities between humans and the rest of the animal kingdom.

As McGrew argues (Ch. 1, this volume), this makes chimpanzees an excellent model for studying the development of complex human behaviour. But, at another level, labelling chimpanzees as culture-bearing merely extends the problem beyond humans to chimpanzees. The central problem is not whether other animals have the capacity for culture, but whether culture is an analytically – as opposed to descriptively – useful concept. In other words, rather than seeking empirical criteria for discerning culture in humans and other animals, we should instead be trying to establish what is actually involved in the evolution of complex behaviour, and how it can be selected for.

The evolution of complex and flexible behaviour

The key characteristic that links humans, chimpanzees and, to a lesser extent, other primates is the complexity and flexibility of their behaviour. It has been argued here that the evolution of this type of behaviour cannot be understood adequately through composite terms such as 'culture', but through considering the components that contribute towards increased behavioural complexity and flexibility. The remainder of this chapter will attempt to construct a model that can account for the evolution of this type of behaviour. However, prior to doing so it is important to establish the basic framework, something that previous models have not always done. In particular, I wish to emphasize the distinction between (a) conditions; (b) the phenotypic characters that are selected; and (c) the epiphenomenal characteristics that arise as a consequence. Conditions constitute the context or selective pressure that prompts or requires complex behaviour. Phenotypic characters are the actual behavioural characteristics that are selected for and come to be incorporated in the behavioural repertoire, resting in the individual and providing it with a reproductive advantage. Epiphenomenal characteristics, which are not themselves selected for, arise as a consequence of the behavioural changes and become part of the selective environment (Table 2.1).

The second important point to establish at the outset is the central role that the brain plays in the evolution of complex behaviour. In strict evolutionary terms, behaviours we observe do not necessarily evolve; what does evolve is the neurological capacity for these behaviours. Primate and human evolution has incorporated a large amount of encephalization, and so a good starting point is to establish the function of the brain in the development of human

Table 2.1 Overall structure of a model to account for the evolution of complex and flexible behaviour patterns.

---

(1) *Conditions for selection*: the ecological and social conditions in which individuals and populations live.

(2) *Evolutionary mechanisms*: principally Darwinian mechanisms of selection operating through an individual's reproductive success (RS). Behaviours that enhance RS under specific conditions will become more frequent within populations.

(3) *Foci for selection*: the acutal phenotypic characteristics that will be selected for. These may be morphological or behavioural.

(4) *The pattern of phenotypic change*: over time there will be phenotypic and genetic changes. Some of these will be empirically observable.

(5) *Epiphenomenal consequences*: these will be the consequences both for the individual and the conditions for selection that occur as a result of evolutionary change.

---

complexity. As far as humans are concerned, it may be argued that the key characteristics of the brain are its propensity for conscious thought and for the use of symbols.

Conscious thought is an analogue for the real world. What goes on inside our heads is the constant construction of models that have some relationship to outside events, be they plans, memories, fantasies, calculations, intentions, etc. In that sense they are very similar to computer simulation programmes, which simulate particular events and processes. I would suggest, as others have done, that encephalization in human evolution is essentially the evolution of a large computer filled with simulation programmes. What has been selected for is the ability to simulate the real world inside our heads, which then results in such complex behaviour patterns. This argument has been developed most fully by Humphrey (1976) in the context of social interactions (predicting, modelling and empathizing with the behaviour of other individuals in a social context), and it has been central to recent discussions of manipulative or 'Machiavellian' primate behaviours (Byrne & Whiten 1986).

This perspective prompts two questions that are essentially separate but often conflated in many discussions: what is the advantage of the ability to simulate the external world (that is, the advantage to the individual that possesses this ability); and what are the conditions likely to promote its evolution (that is, the environmental context in which that advantage occurs)?

The principal advantage of a simulation programme is that it answers 'what if' questions very rapidly and at very low cost and risk. For example, economists can look at the effects of, say, altering tax structures without actually having to carry out the reforms. Simulation programmes enable individuals and institutions to look ahead at the consequences of their decisions and to weigh up alternative courses of action. If the human brain is a computer running simulation programmes, then its advantage will be that an organism can examine inside its head the possible options for responding to the ecological and social problems it faces in its environment. Obviously the advantages that accrue are only as good as the alternatives considered and the viability of the model used (as many economists have found to their, and our, cost), but more appropriate models can be developed with experience.

To return to the original framework, then, when we talk about complex and flexible behaviour, in terms of the phenotypic characters involved, we mean the evolution of the brain into an extremely efficient computer for simulations. As stated at the outset, these attributes reside in the individual and are the focus for selection. The conditions, on the other hand, that promote them are a complex social and ecological environment in which an organism needs to make rapid and flexible responses to the problems it faces (Table 2.2).

From this basic relationship between conditions and selection for phenotypes, and this focus on the evolution of the brain as the genetically controlled generator of behaviour (Table 2.3), arise other characteristics, some of which represent other phenotypic and genetic changes, others of which are more epiphenomenal. Among the former are symbol-use, conscious thought and sensory perception. Symbol-use arises as a consequence of the evolution of the

Table 2.2 Environmental conditions necessary to promote complex and flexible social behaviour.

A. Selective pressures due to particular adaptive problems

a. patchy food supply leads to foraging problems
b. high infant mortality/low reproductive rates lead to problems of allocating parental investment
c. high levels of repeated social interaction lead to problems of achieving a successful reproductive strategy

B. Empirically observable phenomena associated with these selective conditions

a. resource availability and distribution
b. life history parameters
c. social organization (e.g. sexual dimorphism)

Table 2.3 Processes of selection: the brain is the focus for selection for complex behavioural strategies based on the ability to simulate the external world.

A. The brain as a focus for selection

a. simulates solutions to problems faced by organism at low risk/costs to itself
b. allows for novel or unexpected associations that can lead to behavioural innovations

B. Darwinian mechanisms and phenotypic changes

a. the reproductive success of individuals with the best problem-solving brains will be relatively higher
b. there will be direct selection for neurological capacity
c. there will be selection for phenotypes that enhance abilities for making associations and innovating behaviours

C. Empirically observable phenomena

a. brain size and structure
b. rates of behavioural innovation

human brain as a computer simulator. All simulation programmes must be written in a language, and languages are forms of symbols. In terms of the computer analogy pursued here, simulations are likely to have been 'written' in machine code – that is, fundamental neurobiological entities – linked to simple, probably iconic, languages. With continued selection, though, the ability of the brain to handle more and more languages (symbolic systems) and for those languages to become more user friendly (conscious) would increase. These 'languages' may be both internal (thought) or external (for example, speech), and an advantage of this approach means that they can be treated independently. The tremendous capacity for humans to use symbols, both in language and in other systems of internal and external communication, is a consequence of the sophistication of the human brain (Table 2.4).

Another consequence of the evolution of the human brain as a simulator is the evolution of the senses. Any simulation programme is only as good as the information on which it is based. The effectiveness of the simulator will depend, therefore, not just upon the internal efficiency of the brain, but also upon the efficiency with which the brain can receive and process information. In parallel to the evolution of the brain will come an increasing sensitivity in

Table 2.4   Processes of selection that lead to the development of symbolic systems of thought and language as a consequence of the evolution of the brain.

A. Neurological capacities selected for as a result of the development of the brain

a.   coded or symbolic system based on fundamental neurology
b.   increasingly complex and communicable symbolic systems (i.e. consciousness and linguistic skills)

B. Darwinian mechanisms and phenotypic changes

a.   the reproductive success of individuals with symbolic systems capable of coping rapidly with information and communicating it will be selected relatively higher
b.   increasingly there will be selection for interpersonal compatibility of symbolic codes

C. Empirically observable phenomena

a.   brain size and structure
b.   language
c.   art and other symbolic systems
d.   tool form

Table 2.5 Processes of selection: the development of sensory perception in the evolution of human behaviour

---

A. Selection would favour individuals who can acquire the most appropriate quality and quantity of information to enable their brains to function as 'real world' simulators

a.   ability to observe others and the environment
b.   ability to imitate others' behaviour
c.   ability to remember observations
d.   ability to communicate information

B. Darwinian mechanisms and phenotypic changes

a.   the reproductive success of individuals able to link information available to the solution of environmental and social problems will be higher
b.   there will be selection for sensory capabilities and the ability to process information

C. Empirically observable phenomena

a.   within population variability in forms of behaviour

---

certain abilities: the ability to observe what is going on, and to perceive and monitor events; the ability to retain events in the memory; the ability to imitate others; and the ability to put out and receive signs and signals – that is, to communicate. As these are also, like the brain, phenotypic characters, they will become foci for selection. The reproductive success of individuals with these abilities would be higher in the conditions specified earlier (Table 2.5).

Other consequences are more epiphenomenal – that is, they no longer rest with the individual, and therefore cannot be the focus for selection, but arise as a consequence of the behavioural capacity of individuals – and feed back to the individual and its selective environment (Bateson 1988). These include the elaboration of symbolic systems, the degree of intra- and interpopulation variability, the rate of behavioural change, the rate of dissemination of behaviours, and the degree of standardization. These are principally consequences of the ability to innovate behaviours (a response to better simulations) and the improved sensory perceptions (to monitor and adapt to what is going on around an organism). These epiphenomena are what are generally considered to form the basis for cultural evolution (Table 2.6).

Table 2.6 Epiphenomenal consequences of the evolution of complex and flexible behaviour.

A. Expected epiphenomenal characteristics

a.   increased inter- and intra-population variability
b.   increased rates of behavioural change
c.   development of standardized styles and traditions
d.   dissemination through time and space

B. Darwinian mechanisms and phenotypic changes

a.   the selective environment will alter rapidly, favouring individuals able to cope with a dynamic environment

C. Empirically observable phenomena

a.   artefact variability through time and space
b.   patterns of distribution of social groups
c.   variability in social and economic behaviour

Conclusions

In summary, then, the evolution of the capacity of complex and flexible behaviour is accountable in reductionist and Darwinian terms if attention is paid to the distinction between conditions for selection, phenotypic characters, and epiphenomenal characteristics arising as a result. It is stressed that all the components usually considered to constitute culture – for example, those of McGrew & Tutin (1978) – are incorporated in this model (Table 2.7), and there is no need for recourse to higher entities or emergent properties to account for them. Instead, it is argued that analysis should concentrate on the relatively simple component parts and their empirically observable manifestations. As far as palaeoanthropological studies go, culture is a redundant concept except as a linguistic shorthand.

The particular model proposed here is tentative only, but does have several advantages and implications. Principal among these is that by concentrating on the function of the brain it is possible to recognize the essentially mentalistic view of cultural capacity that is central to modern anthropological thinking, rather than its material manifestations. These manifestations, though, are also incorporated, as are also the outcomes of certain selective pressures and phenotypic properties. Another advantage is that the central place given to the distinction between the conditions necessary to promote complex and flexible behaviour and the properties themselves opens up the possibility of investigating the ecological and social context in which these properties evolved (Foley

Table **2.7** Characteristics accounted for by the model presented here (compare McGrew & Tutin 1978).

Complex social and ecological environment	C
Brain expansion and structural development	P1
Use of symbolic codes	P2
Innovation	P2
Imitation	P3
Observation	P3
Perception	P3
Memory	P3
Communication	P3
High rates of behavioural innovation	E
Development of traditions	E
High rates of inter- and intrapopulation variability	E
High rates of information transmission	E

Key: C = Conditions (Table 2.2);
   P1 = phenotypes selected (Table 2.3);
   P2 = phenotypes selected (Table 2.4);
   P3 = phenotypes selected (Table 2.5);
   E = epiphenomenal characteristics observable in the pattern of human behavioural evolution (Table 2.6).

1989). This is particularly important given the current debate concerning the significance of the evolution of anatomically modern humans (Mellars & Stringer 1989). And finally, by focusing on the role of the brain as a means of simulating internally the external world, it may be argued that the development of thought (the internal act of simulation) is independent – and many would argue prior to (Lieberman 1986) – the development of language, the communication of the results of such simulations to others.

It must be stressed that the thrust of this chapter is not to suggest that there are no differences between humans and other animals. There clearly are, and the explanation of these in Darwinian terms remains one of the central problems in evolutionary biology and palaeoanthropology. The intractability of this problem lies in developing models that do not minimize the differences between humans and other species, and yet still use a truly comparative framework. The methodological reductionism of evolutionary biology provides the best scope for this task (Foley 1987). As a summary term, culture perhaps aptly encapsulates many aspects of human uniqueness. However, as generally understood, culture is a synthetic concept, not an analytical one, and as such can have little role to play in the actual investigation of the differences between humans and other forms of life. What is proposed here is not an exact model for the pattern of human behaviour, but a framework for developing such a model, and hence moving towards a truly comparative and evolutionary explanation of human behavioural uniqueness.

References

Bateson, P. P. G. 1988. The active role of behaviour in evolution. In *Evolutionary processes and metaphors*, M. W. Ho & S. W. Fox (eds), 119–207. New York: John Wiley & Sons.
Boesch, C. & H. Boesch 1983. Optimization of nut-cracking with natural hammers by wild chimpanzees. *Behaviour* **83**, 265–86.
Bonner, J. T. 1980. *The evolution of culture in animals*. Princeton: Princeton University Press.
Byrne, R. & A. Whiten 1986. *Machiavellian intelligence*. Oxford: Clarendon Press.
Cavalli Sforza, L. & M. Feldman 1981. *Cultural transmission and evolution*. Princeton: Princeton University Press.
Cheyney, D. L. & R. M. Seyfarth 1980. Vocal recognition in free-ranging vervets. *Animal Behaviour* **28**, 362–7.
Foley, R. 1987. *Another unique species: patterns in human evolutionary ecology*. Harlow: Longman.
Foley, R. 1989. The ecological conditions of speciation: comparative perspectives on the origins of modern humans. In *The human revolution: behavioural and biological perspectives on the origins of modern humans*, P. A. Mellars & C. B. Stringer (eds), 298–320. Edinburgh: Edinburgh University Press.
Gardner, R. A. & B. T. Gardner 1969. Teaching sign language to a chimpanzee. *Science* **165**, 664–72.
Goodall, J. 1970. Tool-using in primates and other vertebrates. In *Advances in the study of behaviour 3*, D. Lehrman, R. Hinde & E. Shaw (eds). New York: Academic Press.
Hinde, R. A. & J. Fisher 1951. Further observations of the opening of milk bottles by birds. *British Birds* **44**, 393–6.
Humphrey, N. K. 1976. The social function of intellect. In *Growing points in ethology*. P. P. G. Bateson & R. A. Hinde (eds), 303–17. Cambridge: Cambridge University Press.
Itani, J. 1958. On the acquisition and propagation of a new food habit in the troop of Japanese monkeys at Takasakiyama. *Primates* **1**, 131–48.
Kroeber, A. L. 1928. Sub-human cultural beginnings. *Quarterly Review of Biology* **3**, 325–42.
Kroeber, A. L. & C. Kluckholm 1952. Culture; a critical review of concepts and definitions. *Papers of the Peabody Museum of American Archaeology and Ethnology* **47**.
Lieberman, P. 1986. *The evolution of language*. Cambridge, Ma.: Harvard University Press.
Lumsden, C. J. & E. O. Wilson 1981. *Genes, mind and culture: the coevolutionary process*. Cambridge, Ma.: Harvard University Press.
McGrew, W. C. 1991. Chimpanzee material culture: what are its limits and why? In *The origins of human behaviour*, R. A. Foley (ed.), Ch. 1. London: Unwin Hyman.
McGrew, W. C. & C. E. G. Tutin 1978. Evidence for a social custom in wild chimpanzees? *Man* **13**, 234–51.
Mellars, P. A. & C. B. Stringer (eds) 1989. *The human revolution: behavioural and*

biological perspectives on the origins of modern humans. Edinburgh: Edinburgh University Press.

Savage-Rumbaugh, E. S., J. L. Pate, J. Lawson, S. T. Smith & S. Rosenbaum 1983. Can a chimpanzee make a statement? *Journal of Experimental Psychology* **112**, 457–92.

Thorpe, W. H. 1961. *Bird Song*. Cambridge: Cambridge University Press.

Wolpoff, M. H. 1971. Competitive exclusion among Lower Pleistocene hominids: the single species hypothesis. *Man* **6**, 601–14.

3 The significance of modern hunter-gatherers in the study of early hominid behaviour

FRANCIS B. MUSONDA

Introduction

The majority of modern hunter-gatherers live mainly in marginal areas – the Kalahari desert, tropical rainforests and the tundra – which are in most cases unsuitable for pastoralism and agriculture. Through specialized adaptations, they have been able to survive in these harsh environments. Their subsistence depends to a large extent on hunting and gathering, but because of differences in habitat characteristics particular activities vary in importance from one region to another.

Certain aspects of our knowledge of modern hunter-gatherers are relevant to the interpretation of the subsistence and sociological behaviour patterns of our early hominid ancestors. In this chapter I focus on four of these: environmental setting, settlement pattern, subsistence activities, and social organization. These are areas which may help to elucidate the cultural dynamics of human evolutionary development.

The fossil evidence itself has so far failed to answer questions relating to the social life of early hominids, adaptive mechanisms that led to bipedal loco-motion, processes of tool-use and tool-making, and subsistence patterns. The last two decades have, however, witnessed a proliferation of ethnographic research on hunter-gatherer and nonhuman primate subsistence patterns and social organization (Lee & DeVore 1968, 1976, Bicchieri 1972, Coon 1971, Silberbauer 1981, Lee 1979, Nelson 1973), and these studies provide an insight into food-gathering strategies and social behaviours that are relevant to the study of early hominids. Much of the data used in this chapter is drawn from studies of hunter-gatherers of eastern and southern Africa, where intensive fieldwork has been undertaken.

Environmental setting

Hunter-gatherer communities on the African continent today are found mainly in three areas: northeastern Zaire, where the Pygmies live; northern Tanzania around Lake Eyasi, where the Hadza are found; and the Kalahari desert in southern Africa, the home of the San. These three areas illustrate the diversity of environmental setting to which these communities have had to adapt.

The region presently occupied by the Hadza is dry, rocky savanna, dominated by thorn scrub and acacia trees (Woodburn 1968, p. 50). Although game is plentiful, vegetable foods constitute about 80 per cent of the intake. In contrast, the Pygmies live in a vast expanse of dense and damp forest which receives abundant rainfall throughout the year (Turnbull 1968, p. 132). They subsist largely on plant foods and on small and medium-sized game (Harako 1981, p. 552). Closely adapted to the forest environment, they continue to practise a hunting and gathering way of life; this adaptation is expressed in their technology and subsistence and is deeply rooted in their ideology (Turnbull 1968, p. 133).

The Kalahari desert, home of the !Kung San, the G/wi San and other hunter-gatherer groups, experiences low rainfall, which varies between 230 and 600 mm a year (Yellen & Lee 1976, Fig. 1.1). As a result these foragers have to cope with extreme scarcity of water throughout the year. Within the Kalahari desert, there exist regional ecological differences which have brought about important shifts in adaptation and cultural and social organization. The central Kalahari, which forms the habitat of the G/wi San is much drier than the Dobe area, where the !Kung are found. Despite being a drier area than the Dobe, the central Kalahari provides a wider spectrum of plant foods which the G/wi are better able to exploit than their !Kung neighbours, for whom the mangetti nuts are a staple. Thus the subsistence pattern of the Kalahari hunter-gatherers is largely dictated by the availability of rainfall, which in turn results in considerable local annual variations in the plant and animal life upon which these hunter-gatherers depend.

Although the environmental settings in which the Hadza, Pygmies, and the San live today are diverse, they are indeed very restricted compared to those of the Plio–Pleistocene hominids of between 3 and 1.5 million years ago. The African hominids during this period inhabited grassland and woodland areas, swamps, and river valleys rather than arid areas or homogeneous expanses of forest. Examples include lake basins and valley floors of major rivers such as the Lower Omo in Ethiopia, East Lake Turkana, and Afar (Leakey 1971, Isaac 1977, Coppens et al. 1976), sites which are lowlands, around 430 m above sea level. These were apparently favoured as habitation areas in eastern Africa by Plio–Pleistocene hominids. The only exception is Laetolil in Tanzania, where footprints and hominid fossils have been found preserved on an upland plain (Leakey et al. 1976) around 1300 m above sea level.

In southern Africa important Plio–Pleistocene hominid localities have been studied. They are all highland sites, ranging from 1161 m above sea level at Taung (Peabody 1964, p. 674), to 1478 m at Sterkfontein, to over 1829 m at Makapansgat (Sampson 1974, p. 18), providing hominid fossil evidence from fissures and caverns. Vrba's (1975) analyses of bovid fossil remains from these hominid sites indicate the presence of open plains and a grassland environment.

Determination of the physical environment in which our early ancestors lived is an essential part of the study of the palaeoecology of early hominids. However, a comparison of modern hunter-gatherer and early hominid environments reveals that these two groups would have required different adaptations because the environments are different. The modern situation is not close enough to that existing in the Plio-Pleistocene period to enable it to play a key part in any reconstruction of the exploitation patterns of our ancestors.

Settlement patterns

The hunter-gatherers of southern and eastern Africa, like others elsewhere in the world, possess tools for their day-to-day economic activities. These largely consist of hunting and gathering tools such as bows and arrows, digging sticks, and an array of domestic tools oriented toward food procurement. There are also ornamental tools and those used for personal hygiene. Simplicity of personal possessions is an advantage to these communities, as they constantly have to abandon camps and set up new ones in places where plant foods and animals are to be found. Since edible vegetable foods vary with the season and the movement of animals depends to a large extent on the availability of plant foods and water, the distribution of hunter-gatherer camps is tailored to meet their need for mobility.

The !Kung and G/wi San live in small, widely scattered camps during the rainy season and aggregate in large camps during the dry season (Yellen 1976), moving frequently within overlapping territories. Territorial boundaries are not well defined or defended. Several groups may move in an area where food resources are available and exploit them together. Usually, these groups are small, their numbers varying according to season, and they have been observed moving campsites from twice to ten times annually (Lee 1976, p. 74). Location of camps is to a large extent determined by the availability of water and food resources. This is especially so with the G/wi San, who move their camps only when they are in search of these resources (Barnard 1979, Hitchcock & Ebert 1984). Thus the shifting of camps is determined by changes in food-procurement strategies and preference, and availability of new food resources or new knowledge about the location of wide-ranging and constantly moving game (Yellen 1976, p. 56).

Similarly, !Kung San settlements tend to be located at least ½ km from a water source in order not to frighten away the game that utilizes it (Yellen 1976). As is true of a G/wi San camp, a !Kung settlement's size and location, and the length of time it is occupied depend on the food resources available to support the group. The placement, spacing, and utilization of work areas and structures in a !Kung camp are influenced by group structure, social organization, and division of activities among different units. These camps are characterized by structural features such as hearths, pits, and tool-manufacture areas, which serve as semipermanent reminders of human habitation. Debris scatter is generally confined to areas surrounding hearths and may consist of vegetable remains, animal bones, and wooden implements. Although most of the usable tools tend to be carried away to new locations when a camp is abandoned, a careful study of debris scatter can make it possible to differentiate between dry- and wet-season camps based on food residue, size of camp (wet-season camps are small, have few huts and occupants, and are briefly inhabited (Yellen 1977, p. 78)), and amount of debris (less in wet- than in dry-season camps). Food resources and length of occupation can be ascertained from the by-products of manufacturing wooden, bone and stone tools. A similar situation in settlement patterning has been observed among the Aborigines of Australia, who live in large camps when water is plentiful and disperse into smaller groups during drier periods of the year (Gould 1969).

The activity patterning in modern hunter-gatherer camps and the settlement pattern evidence in general are important to our understanding of the nature of early hominid sites. Although the ancient eastern and southern African environments were indeed dissimilar from those currently occupied by the San, Hadza, and Pygmies, and although the spatial organization of present-day hunter-gatherers has doubtless been affected by movements of other people and political changes, there are significant archaeological implications to be derived from such studies for the elucidation of early hominid settlements patterns. Schick (1984) has undertaken experiments to study site formation processes and the effects they have upon Palaeolithic archaeological materials in stratified water-laid deposits, especially alluvial sediments. The results have been applied to the study of tool-manufacturing activities at Lower Pleistocene sites at Koobi Fora, Kenya, and have added substantially to our understanding of the behavioural processes involved in the formation of sites and stone tool-manufacture (Toth & Schick 1986).

Subsistence activities

Studies involving modern hunter-gatherers of eastern and southern Africa have shown that these people's survival is largely dependent on their intimate knowledge of the plant and animal communities that they exploit (Lee & DeVore 1968, Marshall 1976, Bicchieri 1972, Tanaka 1976). For example, it has been shown that the !Kung San depend for their survival on the knowledge of places where edible fruits, seeds, roots, bulbs, and other plant foods are to be found and the conditions under which they grow, as well as the feeding habits, movements, and ecological requirements of the game animals upon which they depend. Resources are not uniformly distributed in the Kalahari desert environment, but the San tend to possess extensive knowledge of the environment and this enables them to be self-sufficient in plant and animal foods. During times of plenty plant foods that have a sour taste, are not generally attractive or are inferior in nutrients are not exploited, although these are eaten during periods of food shortages. However, the Dobe area in the Kalahari desert is rich in various vegetable foods throughout most of the year, so the hunter-gatherers here can afford to exercise selectivity in their food quest.

Unfortunately, there are as yet no known Plio-Pleistocene sites which have yielded plant remains suggestive of early hominid diet. However, we do know that the majority of early hominid sites are located close to permanent water sources (Butzer 1978, p. 209, Harris 1980, p. 32, Isaac 1977, Leakey 1971), which may have attracted hominids because of concentration of food resources. A recent study of vegetation transects across east African riparian and nonriparian habitats has found that both diversity and abundance of potentially edible, high-quality plant foods were greatest in riparian habitats, with abundance peaking in the wet season (Sept 1984, 1985). To gain a better understanding of the nature of diet and procurement strategies in the past, archaeological models have to be formulated, based on both contemporary hunter-gatherer subsistence patterns and the results of ecological studies such as the one undertaken by Sept.

Division of labour among hunter-gatherers is based on gender and plays an important part in food acquisition and sharing. Women remain primarily responsible for procuring and preparing vegetable foods, whereas hunting game is largely the responsibility of adult males, although young males and able-bodied females may take part when the need arises. Despite the existence of division of labour among hunter-gatherers, food acquisition remains a collective responsibility. Tasks performed by a hunter-gatherer group become increasingly differentiated with age, with young boys and girls taught to do different kinds of things at an early age (Draper 1975).

An understanding of social aspects relating to division of labour and food-sharing practices may be helpful in elucidating the development of permanent male–female relationships among the early hominids. Based on the fact that both males and females in a modern hunter-gatherer society collect different kinds of foodstuffs, which they then transport back to campsites to share within the social group, an early hominid couple could have paired for the mating season on the basis of food-procurement arrangements. Gradually, a mating system among members of the group would emerge as a result of division of labour and food-sharing practices (see also Lovejoy 1981).

The ethnographic literature on food sharing among hunter-gatherers is not sufficiently detailed to describe how different food items are shared or, in the case of meat, how specific anatomical parts are distributed. What is commonly portrayed is a situation in which every group member receives a share of the available food irrespective of its size, quantity, or nutritional value. However, ethnographic literature provides some insights into the nature of food-procurement strategies and consumption. It has been noted that hunter-gatherers eat some of the food collected almost immediately but also carry some back to the campsites to share with those who stayed behind (Musonda 1986). Food sharing is a characteristic of hunter-gatherers that is deeply entrenched in their eating behaviour (Marshall 1976). However, this behaviour mainly applies to big animals rather than small ones such as tortoises, lizards and duikers. Hunting parties go out to hunt big animals, and meat is shared more or less evenly. According to Marshall (1976, p. 358), when a kill is made hunters eat the liver and other perishable parts on the spot, as well as other body parts, until they are full. The animal is then dismembered and carried back to the campsite. If it is too large to be carried, the bones are left behind after chopping off the meat. Butchering of small animals is carried out at the camp and not necessarily at the kill site. Meat is then shared among members of the group.

This conclusion has important implications for hunter-gatherer responses developed to cope with problems associated with small total calorie and protein intakes. Speth and Spielmann (1983, pp. 18–21) and Speth (1987, p. 17) have discussed some of those responses in relation to modern temperate and northern-latitude hunter-gatherers. When both total calories and protein are in short supply, sharing of meat has to include fats and carbohydrates which are nutritionally significant to prevent hunter-gatherers from losing body weight.

Wilmsen (1978, 1982), Truswell (1977) and Truswell & Hansen (1968) have studied the diet and nutrition of the Kalahari desert hunter-gatherers and have demonstrated that these people undergo significant loss of body weight each year during the late dry season and early rainy season (that is, late spring and early summer). Wilmsen has attributed the weight loss to food shortages in the late dry season and early rainy season. This phenomenon may be applicable to early hominid procurement strategies, because they too may have faced levels of seasonal food stress more or less comparable to the levels faced by contemporary San (Speth 1987, p. 21). According to Speth, the strategies of early hominids towards the procurement of animal protein should be highly dependent on the nutritional status of both hominids and prey, and that nutritional status in turn varies in a systematic fashion with season. Thus the current debate about whether early hominids obtained meat either largely or entirely by hunting or by scavenging (Bunn et al. 1980, Binford 1981, Isaac & Crader 1981, Isaac 1983, 1984, Bunn 1983, Potts 1983, Shipman 1983) may have to look critically at the arguments presented by Speth (1987) concerning the procurement of animal protein during the different seasons of the year.

Subsistence-related behaviour is also reflected in the possession of various kinds of equipment essential to the hunter-gatherer food quest. Studies of toolkits employed by hunter-gatherers are important in our understanding of the economic and social behaviour of early hominids. The fact that almost all the food-acquisition activities of modern hunter-gatherers are accompanied by the use of tools leads us to speculate that a similar kind of behaviour prevailed among early hominids. Ebert's (1979) research among the San of the Kalahari has suggested that certain aspects of tool-use and tool-discard or loss are probably similar to those in the archaeological record, although he admits that the metal knives and axes used by hunter-gatherers today differ in their economic value, effectiveness, and longevity, and in the cultural or symbolic value placed upon them, from the stone implements of earlier hunter-gatherers in the same region. This behaviour relating to tool-use and discard is important to an understanding of past technological remains (Ebert 1979, p. 63).

Studies of modern hunter-gatherers have shown that more than 70 per cent of their food intake consists of plant foods, contrary to the previous emphasis placed on meat-eating and hunting (Ardrey 1961, 1976). Therefore, the primacy of hunting and meat-eating in hominid evolution is not supported by ethnographic studies. Studies of tooth-wear patterns of early hominids suggest a diet that was not dominated by meat (Wallace 1972, Wolpoff 1973), a conclusion that points to the fact that meat-eating was probably not central to hominid origins.

Lower Pleistocene sites in eastern and southern Africa have yielded evidence that points to dependence on a wide range of animal foods by early hominids. Because of preservation problems, no evidence of plant-food gathering has been found at these sites. Today's hunter-gatherers display a broad dietary range, involving a wide spectrum of plant and animal foods, and their intake of these foods ranges from deeply buried tubers to fruits high on trees and from small crawling animals to large mammals. The acquisition of most of these foods is greatly facilitated by the use of tools: wooden spears, bored stones, digging sticks, and bows and arrows.

Modern hunter-gatherers transport meat to campsites in more or less the same fashion that early hominids did, as reflected in the archaeological record from East Lake Turkana (KBS), Kenya, and Olduvai Gorge (FLKN Level 6), Tanzania (Isaac 1976, p. 561, Leakey 1971, p. 252). On the basis of evidence relating to meat-eating, Isaac (1980, p. 226) has argued that the course of human evolution was characterized by a broadening of the range of foods which were important to protohuman ancestral populations. Isaac's argument offers an alternative interpretation to earlier views on human evolution advanced from the 1950s (Dart 1953, Ardrey 1961, 1976, Morris 1967, Jolly 1970) that hunting influenced human evolution and was responsible for the division of labour between the sexes.

Social organization

Models formulated to understand the social behaviour and anatomy of early hominids have largely been based on studies of chimpanzee behaviour and anatomy. These primates are strikingly similar to humans in social behaviour. They prepare and use tools for a variety of purposes, prey upon small animals, occasionally walk bipedally for short distances, share certain foods, and communicate social and environmental information (Goodall 1968, Teleki 1975).

The Pygmy chimpanzee provides an even better fit because this primate is less sexually dimorphic than other apes and is less specialized in habitat, diet, and social behaviour (Zihlman 1979). Zihlman & Tanner (1978, p. 168) have argued that the similarities in behaviour, anatomy, and genes between humans and chimpanzees are so extensive that it is most unlikely that these shared traits are due to convergent evolution.

However, despite the varied activities that chimpanzees are able to perform, their relatively small brain limits their ability to develop highly skilled tool-use and tool-making. Cooperative hunting and food-sharing (plant foods are rarely shared) are evident, but division of labour is not as elaborate as it is among humans. These complex behaviours include carrying food back to a base camp for preparation and sharing, common among contemporary hunter-gatherers.

Studies of modern hunter-gatherers have shown that these communities are characterized by very fluid population distribution over a geographical area. Group structure, like campsite location, is oriented to food and water resources (Yellen & Harpending 1972). Owing to scattered and variable resources, the San constitute a loose confederation of small bands organized through kin and marriage relationships (Silberbauer 1972, p. 273, 1981), and are mobile independent of others in order to achieve a close fit between resource and population density (Yellen & Harpending 1972). The carrying capacity of a territory sets a limit on the size of the band, while the availability of food plants and water is the principal determinant of the band's social organization.

Observations of San bands (Lee 1968, 1976) show that they constitute noncorporate, bilaterally organized groups that live in a single settlement and move together for at least part of the year. Group structure is very variable indeed, perhaps because of changes in rainfall levels and the sparse distribution of standing water in the northern Kalahari.

The social organization of the !Kung is very similar to that of other hunter-gatherers. Damas (1969) has shown that central Eskimos concentrate in large groups in winter when there is good seal-hunting. Also, the Aborigines of Australia follow a concentration-dispersion pattern determined by seasonal differences in water availability. This pattern has also been observed among the Pygmies of the Congo Forest in northeastern Zaire, where the hunter-gatherer movement pattern is based on the seasonal exploitation of key resources and social factors (Turnbull 1965, 1968). According to Lee (1976, p. 91), the existence of this pattern in different kinds of environment suggests that it is basic to the hunting and gathering adaptation. There are indeed several advantages to this kind of pattern: first, a high population density is a distinctive possibility; second, there is a likelihood of responding favourably to the local imbalance in food resources; and third, there is a good chance of keeping the threat of violence to a minimum (Lee 1976, p. 91).

Explanatory models for the social life of early hominids have been drawn from the interpretation of tool-making processes and the way tools were transported. These models are important as they help to define human patterns of behaviour. Evidence from Olduvai Gorge and East Lake Turkana sites has been used to explain how early hominids made stone tools which were carried around and how hunting, the butchering of animals and the sharing of meat were important aspects of social organization (Leakey 1971, Isaac 1978). Language, important for the exchange of information about various aspects of life and a regulating factor of social relations among modern hunter-gatherers, must have been instrumental in the success of an early hominid band. With the development of a mating system and division of labour between sexes, language must have enabled early hominids to develop an alternative 'inheritance', capable of changing faster than genetic systems.

Conclusions

The foregoing is a brief summary of some of the important aspects of modern hunter-gatherer behaviour that palaeoanthropologists are emphasizing in the reconstruction of the cultural history of Plio-Pleistocene hominids (Clark 1968, p. 276). However, opinion remains divided on the question of whether modern hunter-gatherers can be used as exact models for early hominids, especially in view of the former's association with marginal environments. One school of thought argues that a judicious use of ethnographic data may provide a unique opportunity for the reconstruction of the way of life of past populations (Clark 1968, p. 280). A more cautious approach in the use of ethnographic data is advocated by Clark Howell in his contribution at the symposium Man the Hunter (Lee & DeVore 1968, p. 287). He suggests that reconstruction of early hominid life based on the present should be discouraged or very severely curtailed except for very recent time periods. However, later researchers have revealed that some behavioural elements of sociocultural systems have material correlates and can be incorporated in the archaeological interpretation, helping in the making of inferences about early hominid behaviour (Kramer 1979, p. 1).

Yellen's (1977) research among the San has revealed that modern hunter-gatherer societies do provide very significant data for formulating models that are useful in the interpretation of archaeological material. Studies involving the subsistence behaviour of hunter-gatherers point to dependence on both gathering, hunting, and division of labour between sexes, behaviours which were certainly characteristic of early hominids. Hunting, for instance, has been overemphasized as a factor responsible for speeding up human evolution (Washburn & Lancaster 1968, Pfeiffer 1972), whereas vegetable foods have until recently received little attention in discussions related to human evolution. Studies of hunter-gatherers show that meat is a minor component in their diet (between 20 and 50 per cent), so in the light of this information hunting cannot be regarded as a factor responsible for human development.

Although modern hunter-gatherer studies have made it possible for archaeologists to speculate on the size of early hominid social groups, the length of time involved in refuse accumulation, subsistence, and settlement patterns, serious misgivings must remain about developing models based on present-day hunter-gatherer activities. Modern groups are far removed in time from the early hominids. To use them to postulate past activities is to suggest that the subsistence base and technology have not changed since the Plio-Pleistocene. While such studies are undeniably useful, it is important to realize that modern hunter-gatherers inhabiting marginal areas may differ from prehistoric peoples inhabiting different environments. Moreover, in the course of time these hunter-gatherers may have undergone considerable

change (Tanaka 1976), requiring different adaptations. Thus the use of ethnographic analogy in the interpretation of archaeological data that are greatly
removed in time and space is risky, to say the least (Binford 1968, Isaac 1972),
and may have only limited application. It should not be assumed that the
observed differences between agricultural and pastoral societies, on the one
hand, and hunter-gatherers, on the other, are an indication of the closeness of
the latter to the Plio-Pleistocene hominids. However, as long as contemporary
hunter-gatherers are not viewed as 'living fossils' surviving from more or less
remote periods (Isaac 1968, p. 253), prehistoric studies can use the insights they
provide to devise research in the archaeological context (Isaac 1968, 1972,
p. 172, Clark 1968). Such an approach offers unique opportunities for the
reconstruction of early hominid activities in the distant past.

Acknowledgement

I am greatly indebted to Florence Nchimunya, of the Livingstone Museum, who
typed the draft of this chapter.

References

Ardrey, R. 1961. *African genesis*. New York: Collins.
Ardrey, R. 1976. *The hunting hypothesis*. New York: Atheneum.
Barnard, A. 1979. Kalahari bushmen settlement patterns. In *Social and ecological systems*,
    P. Burnham & R. F. Ellen (eds), 131–44. London: Academic Press.
Bicchieri, M. (ed.) 1972. *Hunters and gatherers today*. New York: Holt, Rinehart &
    Winston.
Binford, L. R. 1968. Post-Pleistocene adaptations. In *New perspectives in archaeology*,
    S. R. Binford & L. R. Binford (eds), 313–41. Chicago: Aldine.
Binford, L. R. 1981. *Bones: ancient men and modern myths*. New York: Academic Press.
Bunn, H. T. 1983. Evidence on the diet and subsistence patterns of Plio-Pleistocene
    hominids at Koobi Fora, Kenya, and Olduvai Gorge, Tanzania. In *Animals and
    archaeology 1, Hunters and their prey*, J. Clutton-Brock & C. Grigson (eds), 21–30.
    BAR International Series 163. Oxford: British Archaeological Reports.
Bunn, H. T., J. W. K. Harris, G. Isaac, Z. Kaufulu, E. Kroll, K. Schick, N. Toth &
    A. K. Behrensmeyer 1980. FxJj50: an early Pleistocene site in northern Kenya.
    World Archaeology 12, 109–36.
Butzer, K. W. 1978. Geological perspectives on early hominid evolution. In *Early
    hominids of Africa*, C. J. Jolly (ed.), 191–217. New York: St Martin's Press.
Clark, J. D. 1968. Studies of hunter-gatherers as an aid to the interpretation of
    prehistoric societies. In *Man the hunter*, R. B. Lee & I. DeVore (eds), 276–80.
    Chicago: Aldine.
Coon, C. S. 1971. *The hunting peoples*. Boston: Little, Brown.
Coppens, Y., F. C. Howell, G. Isaac & R. E. F. Leakey (eds) 1976. *Earliest man and
    environments in the Lake Rudolf Basin*. Chicago: University of Chicago Press.
Damas, D. 1969. Characteristics of central Eskimo band structure. In *Contributions to
    anthropology: band societies*, D. Damas (ed.), 116–38. National Museum of Canada
    Bulletin. 228.

Dart, R. A. 1953. The predatory transition from ape to man. *International Anthropological and Linguistic Review* **1** (4), 201–19.

Draper, P. 1975. !Kung women: contrasts in sexual egalitarianism in the foraging and sedentary contexts. In *Toward an anthropology of women*, R. Reiter (ed). New York: Monthly Review Press.

Ebert, J. I. 1979. An ethnoarcheological approach to reassessing the meaning of variability in stone tool assemblages. In *Ethnoarcheology: implications of ethnography for archeology*, C. Kramer (ed.), 59–74. New York: Columbia University Press.

Goodall, J. 1968. The behaviour of free-living chimpanzees in the Gombe Stream Reserve. *Animal Behaviour Monographs* **1**, 165–311.

Gould, R. A. 1969. Subsistence behaviour among the Western Desert Aborigines of Australia. *Oceania* **39**, 253–74.

Harako, R. 1981. The cultural ecology of hunting behaviour among Mbuti Pygmies in the Ituri Forest, Zaire. In *Omnivorous primates: gathering and hunting in human evolution*, R. S. O. Harding & G. Teleki (eds), 499–555. New York: Columbia University Press.

Harris, D. R. 1980. Commentary: human occupation and exploitation of savanna environments. In *Human ecology in savanna environments*, D. R. Harris (ed.), 31–39. London: Academic Press.

Hitchcock, R. K. & J. I. Ebert 1984. Foraging and food production among Kalahari hunter-gatherers. In *From hunters to farmers: the causes and consequences of food production*, J. D. Clark & S. A. Brandt (eds), 328–48. Berkeley: University of California Press.

Isaac, G. L. 1968. Traces of Pleistocene hunters: an East African example. In *Man the hunter*, R. B. Lee & I. DeVore (eds), 253–61. Chicago: Aldine.

Isaac, G. L. 1972. Early phases of human behaviour: models in Lower Palaeolithic archaeology. In *Models in archaeology*, D. L. Clarke (ed.), 167–99. London: Methuen.

Isaac, G. L. 1976. The activities of early African hominids: a review of archaeological evidence from the time span two and a half to one million years ago. In *Human origins: Louis Leakey and the East African evidence*, G. L. Isaac & T. McCown (eds), 462–514. Menlo Park, California: W. A. Benjamin, Inc.

Isaac, G. L. 1977. *Olorgesailie: archaeological studies of a Middle Pleistocene lake basin in Kenya*. Chicago: University of Chicago Press.

Isaac, G. L. 1978. The foodsharing behaviour of protohuman hominids. *Scientific American* **238** (4), 110–23.

Isaac, G. L. 1980. Casting the net wide: a review of archaeological evidence for early hominid land use and ecological relations. In *Current argument on early man*, L. Konigsson (ed.), 226–51. Oxford: Pergamon Press.

Isaac, G. L. 1983. Bones in contention: competing explanations for the juxtaposition of early Pleistocene artefacts and faunal remains. In *Animals and archaeology 1. Hunters and their prey*. J. Clutton-Brock & C. Grigson (eds), 3–19. BAR International Series 163. Oxford: British Archaeological Reports.

Isaac, G. L. 1984. The archaeology of human origins: studies of the Lower Pleistocene in East Africa 1971–1981. *Advances in World Archaeology* **3**, 1–87.

Isaac, G. L. & D. C. Crader 1981. To what extent were early hominids carnivorous? In *Omnivorous primates*, R. S. O. Harding & G. Teleki (eds), 37–103. New York: Columbia University Press.

Jolly, C. 1970. The seed-eaters: a new model of hominid differentiation based on a baboon analogy. *Man* **5** (1), 5–26.

Kramer, C. (ed.) 1979. *Ethnoarcheology: implications of ethnography for archeology*. New York: Columbia University Press.

Leakey, M. D. 1971. *Olduvai Gorge Volume 3*. Cambridge: Cambridge University Press.

Leakey, M. D., R. L. Hay, G. H. Curtis, R. E. Drake, M. K. Jackes & T. D. White 1976. Fossil hominids from the Laetolil beds. *Nature* **262**, 460–66.

Lee, R. B. 1968. What hunters do for a living, or how to make out on scarce resources. In *Man the hunter*, R. B. Lee & I. DeVore (eds), 30–48. Chicago: Aldine.

Lee, R. B. 1976. !Kung spatial organization: an ecological and historical perspective. In *Kalahari hunter-gatherers: studies of the !Kung San and their neighbors*, R. B. Lee & I. DeVore (eds), 74–97. Cambridge, Ma.: Harvard University Press.

Lee, R. B. 1979. *The !Kung San: men, women and work in a foraging society*. Cambridge: Cambridge University Press.

Lee, R. B. & I. DeVore (eds) 1968. *Man the hunter*. Chicago: Aldine.

Lee, R. B. & I. DeVore 1976. *Kalahari hunter-gatherers: studies of the !Kung San and their neighbors*. Cambridge, Ma.: Harvard University Press.

Lovejoy, C. O. 1981. The origin of man. *Science* **211**, 341–50.

Marshall, L. 1976. Sharing, talking and giving: relief of social tensions among the !Kung. In *Kalahari hunter-gatherers: studies of the !Kung San and their neighbors*, R. B. Lee & I. DeVore (eds), 349–71. Cambridge, Ma.: Harvard University Press.

Morris, D. 1976. *The naked ape*. London: Jonathan Cape.

Musonda, F. B. 1986. Plant food in the diet of the prehistoric inhabitants of the Lunsemfwa drainage basin, Zambia, during the last 20 000 years. *Zambia Geographical Journal* **36**, 17–27.

Nelson, R. K. 1973. *Hunters of the northern forest*. Chicago: University of Chicago Press.

Peabody, F. E. 1964. Travertines and cave deposits of the Kaap escarpment of South Africa, and the type locality of *Australopithecus africanus* Dart 1924. *Bulletin of the Geological Society of America* **65**, 671–706.

Pfeiffer, J. 1972. *The emergence of man*. New York: Harper.

Potts, R. 1983. Foraging for faunal resources by early hominids at Olduvai Gorge, Tanzania. In *Animals and archaeology 1. Hunters and their prey*, J. Clutton-Brock & C. Grigson (eds), 51–62. BAR International Series 163. Oxford: British Archaeological Reports.

Sampson, C. G. 1974. *The Stone Age archaeology of southern Africa*. New York: Academic Press.

Schick, K. D. 1984. *Processes of Palaeolithic site formation: an experimental study*. Unpublished PhD thesis, University of California, Berkeley.

Sept, J. M. 1984. *Plants and early hominids in east Africa: a study of vegetation in situations comparable to early archeological site locations*. Unpublished PhD thesis, Department of Anthropology, University of California, Berkeley.

Sept, J. M. 1985. Eden's forbidden fruit? Plant food foraging opportunities in east African habitats. Paper presented at the 50th Annual Meeting of the Society for American Archaeology, Denver, Col. (4 May 1985).

Shipman, P. 1983. Early hominid lifestyle: hunting and gathering or foraging and scavenging? In *Animals and archaeology 1. Hunters and their prey*, J. Clutton-Brock & C. Grigson (eds), 31–50. BAR International Series 163. Oxford: British Archaeological Reports.

Silberbauer, G. B. 1972. The G/wi Bushmen. In *Hunters and gatherers today*, M. Bicchieri (ed.), 271–325. New York: Holt, Rinehart & Winston.

Silberbauer, G. B. 1981. *Hunter and habitat in the central Kalahari desert*. Cambridge: Cambridge University Press.

Speth, J. D. 1987. Early hominid subsistence strategies in seasonal habitats. *Journal of Archaeological Science* **14**, 13–29.

Speth, J. D. & K. Spielmann 1983. Energy source, protein metabolism and hunter-gatherer subsistence strategies. *Journal of Anthropological Archaeology* **2**, 1–31.

Tanaka, J. 1976. Subsistence ecology of central Kalahari San. In *Kalahari hunter-gatherers: studies of the !Kung San and their neighbors*, R. B. Lee & I. DeVore (eds), 98–119. Cambridge, Ma.: Harvard University Press.

Teleki, G. 1975. Primate subsistence patterns: collector-predator and gatherer-hunter. *Journal of Human Evolution* **4**, 125–84.

Toth, N. & K. D. Schick 1986. The first million years: the archeology of protohuman culture. In *Advances in archeological method and theory, Volume 9*, M. B. Schiffer (ed.), 1–96. New York: Academic Press.

Truswell, A. S. 1977. Diet and nutrition of hunter-gatherers. In *Health and disease in tribal societies*, 213–26. Ciba Foundation Symposium 49. Amsterdam: Elsevier.

Truswell, A. S. & J. D. L. Hansen 1968. Medical and nutritional studies of !Kung bushmen in northwest Botswana: a preliminary report. *South African Medical Journal* **42**, 1338–9.

Turnbull, C. 1965. *Wayward servants: the two worlds of the African Pygmies*. Garden City: Natural History Press.

Turnbull, C. 1968. The importance of flux in two hunting societies. In *Man the hunter*, R. B. Lee & I. DeVore (eds), 132–37. Chicago: Aldine.

Vrba, E. 1975. Some evidence of chronology and palaeoecology of Sterkfontein, Swartkrans and Kromdraai from the fossil Bovidae. *Nature* **254**, 301–4.

Wallace, J. A. 1972. Tooth chipping in the australopithecines. *Nature* **244**, 117–18.

Washburn, S. L. & C. S. Lancaster 1968. The evolution of hunting. In *Man the hunter*, R. B. Lee & I. DeVore (eds), 293–303. Chicago: Aldine.

Wilmsen, E. N. 1978. Seasonal effects of dietary intake on Kalahari San. *Federation Proceedings* **37**, 65–72.

Wilmsen, E. N. 1982. Studies in diet, nutrition, and fertility among a group of Kalahari bushmen in Botswana. *Social Science Information* (Sage: London and Beverly Hills) **21** (1), 95–125.

Wolpoff, M. H. 1973. Posterior tooth size, body size and diet in South African gracile australopithecines. *American Journal of Physical Anthropology* **39**, 375–94.

Woodburn, J. 1968. An introduction to Hadza ecology. In *Man the hunter*, R. B. Lee & I. DeVore (eds), 49–55. Chicago: Aldine.

Yellen, J. E. 1976. Settlement patterns of the !Kung: an archeological perspective. In *Kalahari hunter-gatherers: studies of the !Kung San and their neighbors*, R. B. Lee & I. DeVore (eds), 47–72. Cambridge, Ma.: Harvard University Press.

Yellen, J. E. 1977. *Archeological approaches to the present: Model for reconstructing the past*. New York: Academic Press.

Yellen, J. E. & Harpending 1972. Hunter-gatherer populations and archaeological inference. *World Archaeology* **4** (2), 244–53.

Yellen, J. E. & R. B. Lee 1976. The Dobe-/Du/da environment: background to a hunting and gathering way of life. In *Kalahari hunter-gatherers: studies of the !Kung San and their neighbors*, R. B. Lee & I. DeVore (eds), 27–46. Cambridge, Ma.: Harvard University Press.

Zihlman, A. 1979. Pygmy chimpanzee morphology and the interpretation of early hominids. *South African Journal of Science* **75**, 163–5.

Zihlman, A. & N. Tanner 1978. Gathering and the hominid adaptation. In *Female hierarchies*, L. Tiger & H. Fowler (eds). Chicago: Fowler AVC, Inc.

4 *Archaeological evidence for modern intelligence*

THOMAS WYNN

Introduction

Many prehistorians assume that the evolution of anatomically modern humans coincided with the appearance of certain attributes of behavioural complexity in the archaeological record: parietal art, exchange systems, and curated tools to name just a few. Some argue that this complexity reflects a more powerful intelligence and that *Homo sapiens sapiens* was blessed with a cleverness that gave him a marked advantage over his archaic predecessors. According to Redman, there was 'a change in adaptive strategies and organizational abilities at the beginning of the Upper Palaeolithic. This transition signifies the rapidly increasing ability of human beings to recognize environmental potentials that existed [and] to communicate these potentials to others' (1978, pp. 51–2). In a discussion of one aspect of complexity, storage, Binford makes the following contention: 'It is my impression that the *ability to anticipate events and conditions not yet experienced* was not one of the strengths of our ancestors prior to the appearance of clear evidence for symboling, e.g. personal ornaments, graphics in the form of painting, 'art', and notation ...' (1982, p. 178, emphasis in original). In other words, prior to the Upper Palaeolithic, *Homo* was incapable of planning very far ahead. Both the supposed foresight of *H. sapiens sapiens* and his increased organizational ability, if true, must be aspects of a more powerful intelligence.

The question of intelligence is not a matter of sophistry. If the behavioural complexity we see in the archaeological record was tied to intelligence, then we must incorporate a factor of biological evolution into our interpretation of culture change (see Foley, Ch. 2, this volume). Intelligence, at least as commonly conceived, has a physiological component that must have evolved. If, on the other hand, this complexity was not tied to intelligence, then we must interpret culture change rather differently. These two alternatives constitute very different understandings of the nature of later human evolution.

In this chapter I address the question of the appearance of modern intelligence using the theory of Jean Piaget, perhaps the most influential developmental psychologist of the 20th century. For evidence I use the archaeological record. Most of my examples come from the European Upper Palaeolithic, not because it is somehow more typical but because the transition to increased cultural complexity was relatively abrupt. In particular I discuss the significance of the following behaviours:

1 technology, especially curated tools and facilities;
2 subsistence, especially seasonal hunting and fishing;
3 exchange systems;
4 ritual systems, especially Magdalenian parietal art.

Archaeology and intelligence

In any study of intelligence we immediately run into some methodological problems. The first is definition. In introductory psychology courses, one is taught that intelligence is something measured by IQ tests. In other words, it is performance on a standardized test. This definition clearly has very little evolutionary potential; we cannot give Neanderthals the Stanford-Binet. Furthermore, saying that Johnny has a higher IQ than Tommy is not quite the same as saying that elephants are more intelligent than monitor lizards. Intelligence is too general and fuzzy a concept to use without narrowing it down a bit. The second problem is one of evidence. What do we look for? Archaeology must, with few exceptions, rely on the analysis of the end products of behaviour. Some of these may have required more intelligence than others, but if so it is not obvious which. How do we select our attributes? We can solve both of these methodological difficulties if we turn to established theories of intelligence.

Unfortunately, archaeologists more often than not turn to common sense. We consider ourselves to be intelligent people and, by self-reflection, we decide what it is in prehistory that should require a high intelligence. Self-reflection is a notoriously faulty source for scientific concepts. This problem is not restricted to intelligence. Binford (various: see, for example, 1983) has shown that most faunal analysis has been based on common-sense ideas that are simply wrong. He has attempted to replace the common sense with experimentally based 'middle-range theory', as he terms it. One problem with common-sense ideas of intelligence is their tendency to confuse intelligence with complexity: there are more tool types in the Upper Palaeolithic, therefore people must have been smarter. Pursuing such reasoning, one would in turn have to argue that 20th-century Europeans are more intelligent than 19th-century Europeans. There are more insidious dangers. Based largely on self-reflection and common sense, 19th-century scientists assumed that men were smarter than women and northern Europeans were smarter than southern Europeans. After all, it seemed obvious. Worse still, this bias led them to find (or make!) measures corroborating their ideas (Gould 1981). Common sense is simply insufficient. Luckily for the study of intelligence, well-developed and well-tested theories exist and, unlike Binford, we need not construct our own experimental base.

For a theory of intelligence to be useful for the archaeologist it must do two things. First, it must define intelligence in such a way that it encompasses the behaviours of nonhumans. It must see intelligence as an entity that varies from taxon to taxon and which can evolve within a single lineage. It must be able to compare elephants to monitor lizards, apes to humans, and then to measure the differences in some way. The IQ definition fails here. Second, the theory must be able to assess the end products of behaviour. Many theories are based on the assessment of sequences of behaviour or on verbal accounts by subjects. We have neither in the archaeological record. A third characteristic is also important. The theory must be persuasive. The categories of intelligence defined by the theory need to have been confirmed again and again in comparative studies and cross-cultural studies. The truth and reliability of the theory must be established on contemporary data. Only then can the theory be applied to prehistory. The archaeological record does not have the resolution to generate and test theories of intelligence on its own.

One methodological caveat cross-cuts all theories of intelligence used by archaeologists: the problem of minimum necessary competence. We cannot assume that the behaviours we see represented in archaeological evidence required the highest abilities of the prehistoric people. They may well have used very sophisticated thinking in domains that are archaeologically invisible – social structure or cosmogony, for example. But when we assess intelligence archaeologically, we can reach conclusions only about the minimum competence necessary for the behaviour that we see. It is therefore possible to underestimate intelligence, especially because archaeological evidence consists mostly of mundane day-to-day behaviours that may not have taxed prehistoric intelligence, just as they do not tax ours. On the other hand, it would be very difficult to *over*estimate intelligence since we must assess minimum abilities.

A Piagetian approach to prehistoric intelligence

Piaget's genetic epistemology is well known as a theory of child development, a theory that describes a sequence of stages through which all children pass, from infancy to adolescence. But Piaget considered the theory to be much more general and, indeed, intended that it should describe the development of all forms of knowing, from the evolution of intelligence to the history of scientific thinking (Piaget 1970, 1972). He studied human children because they present a readily available sequence of development, not because he was interested in education or in child-rearing. Piaget's theory has been extensively applied in studies of development, including cross-cultural and interspecific studies. It is probably the most widely applied and closely examined theory of intelligence yet devised.

Piagetian theory is a structural theory that defines intelligence as organizational ability. This encompasses the way an organism places itself in, moves about in, and manipulates its surroundings. The theory makes very specific predictions about the form a child will use to solve a particular kind of problem and, while the theory was not designed to assess results, many of the forms of organization it describes can be used to evaluate the products of behaviour. The theory, then, meets the criteria already outlined: it can be used to compare nonhumans and to assess products of behaviour.

The theory is also a stage theory. The stages were described on the basis of observations of children's approaches to tasks, ways of solving problems, and so on. Piaget's scheme includes four major stages – sensorimotor, preoperational, concrete operational, and formal operational – and each of the stages includes substages. The sequence is invariant. Every child passes through the stages and substages in the same order, though the age at which each stage is achieved varies from child to child. The defining criteria are qualitative and not based on statistical trends. An important part of most Piagetian experiments is a dialogue between the experimenter and the child, for a child's reasons for behaving in a particular way are as enlightening as the products themselves. The dialogue aspects cannot, of course, be applied in prehistory but the typical products of certain kinds of organization can be used, though the precision of the analysis is reduced. One kind of behaviour used often by Piaget is spatial ability: arranging objects, drawing figures, reconstructing scenes, mapping, and so on. It is this emphasis on spatial ability that allows us to use the scheme as a yardstick in prehistory.

As important as the stages are to Piagetian theory, they are not its core. This lies in Piaget's view of the nature of intelligence and the process of development. It is a structural theory but does not view structure as innate. Rather, structure is *constructed* by means of an interaction between individuals and their environment. Individuals apply their internal organization to the external context of their surroundings and, if their organization is inadequate, modify the internal structure based on this experience. It is an active construction of a new organization, not a behaviourist kind of passive learning. The new organization is in turn applied until it is inadequate, then modified, and so on. The result is a sequence of more and more powerful organizations that are expressed behaviourally as the stages. The theory is not innatist, like Chomskian structuralism, nor is it behaviourist, like Skinnerian psychology. However, it is cognitive in the sense that the brain actively constructs the organizations. Piaget is not mystical about the structures but sees them as being the manifestations of a brain organized by its own action.

While Piaget's idea of constructivism is most easily understood in the context of ontogeny, he intended it to apply to all development, including phylogenetic development. Piaget's first publications were in biology at a time when Haeckel's idea of recapitulation was still influential and while Piaget never argued for a strict 'terminal addition with acceleration' (Gould 1977), he always maintained that the ontogenetic sequence informed us about the phylogenetic sequence. The constructivist nature of development accounts for the parallel. The structure typical of one stage is a logically necessary prerequisite for the next in that the succeeding stage builds on and out of the organization of its antecedent. This logical necessity must be true of any sequence, including both ontogeny and phylogeny. Piaget himself did not pursue prehistory, although he occasionally mentioned it, and was content to study his invariant sequence in its most accessible form. Even if we do not share his certainty about the parallel, the scheme still provides a powerful hypothesis for the phylogenetic sequence. This is the approach I will take.

Because I am concerned here with the appearance of anatomically modern humans, I will deal only with Piaget's final stages of operational intelligence. Elsewhere (Wynn 1981, 1985, 1989) I have considered preoperational stages. There are two organizational features that are central to operational thinking but which do not appear in earlier stages: reversibility and conservation. A simple example of reversibility is in arithmetic, where every operation has an implied inverse: for example, addition is the inverse of subtraction. Conservation is one of Piaget's most famous concepts. In transitivity when $A = B$ and $B = C$ it must follow that $A = C$. Something has been conserved across the relationship. A preoperational child does not see any logical necessity in transitivity and insists that A and C must be directly compared before the answer can be known. Reversibility and conservation provide thinking with some very useful organizational features. One is precorrection of errors. 'What this means is that an operational system is one which excludes errors before they are made, because every operation has its inverse in the system . . .' (Piaget 1970, p. 15). With operational thinking an individual can make detailed contingency plans by, in a sense, returning to a starting point in thought (reversibility) after anticipating possible difficulties. The preoperational thinker can proceed only by trial and error because reversibility in his or her planning is lacking. Reversibility and conservation also allow classification. Classification requires reversibility (subclass A + subclass $A' = B$; class B − subclass A' = subclass A) and the conservation of some definitional variable across disparate items or groups. Preoperational thought can group accordingly to similarity but cannot create logically consistent classifications or reclassifications. From this bare description I hope it is clear that operational thinking is indispensable to many kinds of human behaviour − complex kinship systems and interplanetary exploration to name two. It must have evolved, but when?

Concrete operations

Operational structures do not emerge overnight in ontogeny. Piaget's scheme is often caricatured as if this were the case, but he never suggested that the transition from one kind of thinking to another occurred in a single flash of insight. Rather, operational thinking is first applied in a narrow range of domains and then applied to more and more situations. For example children can conserve quantity before they can conserve weight. However, the scheme does entail a development within operational thinking. This is the distinction between concrete operations and formal operations. Piaget considers that these styles of thinking constitute two separate stages. Formal operations are the final achievement of adult intelligence and contain abstract features not found in concrete operations.

Concrete operations are characterized by all of the organizational features of operations: reversibility, conservation, precorrection of errors, and so on. They are the first operations to appear and are used to organize tangible things, like objects and people, and simple concepts, like numbers – hence the term concrete. Hypothetical entities or abstract concepts are not the stuff of concrete operations. Using concrete operations, one can classify objects according to colour and reclassify them according to shape, but cannot then hypothesize about the class of all classes. One accepts the necessity of division as the inverse of multiplication, but sees no necessity in the square root of minus one. Concrete operations are, nevertheless, a powerful organizational tool; indeed, they are the principal organizational tool for day-to-day living. Tasks, tools, kinship, politics, and religion are all organized in this manner. Concrete operations have been documented again and again in cross-cultural contexts (Dasen 1977, Dasen & Herron 1981). The cross-cultural use of Piaget's theory is fraught with methodological problems (see p. 63), but it does appear that the stage sequence is the same for all groups and that concrete operations are achieved by modern adults everywhere.

We can now turn to the archaeological record.

Archaeological evidence for concrete operations

There is good evidence for the use of concrete operations by 300 000 years ago. I have presented this argument in detail elsewhere (Wynn 1979, 1989) but a precis is appropriate here.

One of the advantages of Piagetian theory for prehistory is its emphasis on spatial relations. Archaeologists have stone tools in abundance and many of these present patterns that can be used to infer the minimum spatial competence of the stone knapper. By the end of the Acheulean, and perhaps a bit earlier, there were stone tools of considerable spatial sophistication. The one familiar to most prehistorians is the fine handaxe with true bilateral symmetry and lenticular cross-section. True symmetry is a Euclidean relation that is achieved only in the concrete operational stage (Piaget & Inhelder 1967). The mirroring of a shape across a midline requires reversibility, because the shape must be inverted in thought. It cannot have been achieved by trial-and-error copying because the stone could not be folded to compare one side to the other (as is done in the symmetry of paper dolls). I must emphasize that I am referring only to those handaxes that demonstrate an almost perfect symmetry, and one that was the result of extensive trimming. These demand that the knapper have a concept of symmetry. Most handaxes, indeed probably all of the early ones, are only roughly symmetrical and there are ways to achieve this without a symmetry concept (Wynn 1985). The lenticular cross-sections of fine handaxes are even more demanding. These are symmetrical figures but, more importantly, they cannot be directly perceived by the knapper. They must be constructed in thought. Especially fine handaxes have a virtual infinity of symmetrical cross-sections, all of which the knapper must have considered while trimming the piece. Such a feat is beyond the ability of preoperational trial-and-error plans, which can consider only one variable at a time. By the end of the Acheulean there are also minimally trimmed handaxes, which achieve a remarkable symmetry with very little trimming. These suggest a sophisticated idea of the relation of whole to parts, in this case the relation of short trimming segments to the conceived final product. Again, such a concept requires reversibility and precorrection of errors and is beyond the scope of trial-and-error plans.

The spatial evidence from stone-tool geometry may seem meagre but it is actually quite compelling. At least three different spatial relations that were used required operational structures. These spatial patterns cannot be produced by preoperational organizations. We must therefore conclude that the minimum competence of these later Acheulean stone knappers was concrete operational intelligence.

Formal operations

The structures of formal operational thinking are more generally applied than those of concrete operations. No longer is the logic applied only to objects or to real data sets; it is used to establish generalities about all possible situations. This development also includes the capacity for hypotheticodeductive reasoning, the use of propositional logic, and the ability to disassociate form from content. In other words, formal operations are characteristic of the most sophisticated kind of reasoning we know. It is the final stage of Piaget's scheme and also the most controversial. I will here investigate the possibility that formal operations were associated with the appearance of anatomically modern humans (*Homo sapiens sapiens*) and that this development supplied them with some advantage.

In addition to the general claims for hypothetical reasoning and so on, Piaget (Inhelder & Piaget 1958) argues for a very specific change in the logic of formal operations. While concrete operations employ reversibility, formal operations coordinate two kinds of reversibility: inversion and reciprocity. In inversion a transformation is combined with an inverse that negates the transformation: for example, $+A - A = 0$. This is the kind of reversibility used in the classification example above ($A + A' = B$, $B - A' = A$) and also in the whole–part relations demonstrated by minimally trimmed handaxes, where the addition or elimination of potential trimming segments *in thought* is a matter of inversion (see Wynn 1979 for detailed argument). Reciprocity is simply a reversal of order (Piaget 1970, p. 22). A transformation combined with its reciprocal yields an equivalence (as opposed to negation): for example, $A \leqslant B$ combined with its reciprocal $B \leqslant A$ results in $B = A$. This kind of reversibility is beautifully exemplified by the fine bilateral symmetry of the later handaxes, where the shape is mirrored by its reciprocal. While the handaxes demonstrate both kinds of reversibility, we cannot argue that they were coordinated into a formal system. Such a coordinated system has some interesting properties and an example from Piaget's work is in order.

Understanding the relation of weight to distance on a balance scale requires an understanding of proportion, which here requires coordinating inversion and reciprocity. An individual using formal operations knows that a balance can be achieved by adding and subtracting weight (inversion), moving the weights in or out on the arms (reciprocity), *or* by adding weight to one arm and moving a smaller weight further out on the other (a coordination of the two). After only a brief experimentation the formal operational thinker can generalize the proportions to all possible situations. Individuals using concrete operations can balance by adding weight or by moving weight but do not construct a system of proportions that they see as being always and everywhere true. It is not that concrete operational individuals cannot balance the weights, only that they have no foolproof system.

Archaeological evidence for formal operations

Ideally, the arachaeologist would look for evidence of the system of coordinated reversibilities since this is the most specific difference between formal and concrete operations. Unfortunately, such evidence is not easy to find and, in the absence of texts, may well be impossible in prehistory. The difference between concrete and formal operations is not so much in the end product as in the way the solution is achieved. After all, concrete operations can balance the scale but archaeologists would find only the balanced scale, not the thought process behind it. Formal operations, generally considered, are not used on tangible things but on hypotheses, generalizations, and contentless forms. Unlike concrete operations, they will not be directly preserved in patterns whose minimum necessary competence is formal operations. Patterns of objects demand at most concrete operations. As a consequence the archaeologist must take one step beyond the physical evidence and assess competence based on *interpretations* of prehistoric behaviour.

In the following analysis I will focus on selected examples from four domains of behaviour: technology, subsistence, social organization, and ritual and art.

Although technology alone was sufficient to document concrete operational thinking at 300 000 years ago, it is of little help in documenting formal operations. Stone tools in particular are uninformative, even though they are the most abundant source of palaeolithic evidence. Much has been written about the sophistication of such techniques as the Levallois and prismatic blade core, but they are no more conceptually difficult than the fine bifaces of the later Acheulean. They may require more skill and practice (though this is debatable), but the minimum conceptual requirement is reversibility in monitoring the relation of core and flakes (whole–part relations) and precorrection of errors (Wynn 1985). Nothing in stone knapping requires the coordinated systems of formal operational intelligence. Prismatic cores may have made more efficient use of raw material but efficiency is not necessarily a mark of intelligence.

Two post-Acheulean developments in technology are provocative: curated tools and the use of facilities. Again, both curation and the manner in which facilities were used is a matter of interpretation, not a pattern directly observed. Curated tools are not manufactured for a specific task but have a general function and are used again and again, carried from place to place. Elements of Upper Palaeolithic technology were almost certainly curated, Magdalenian bone points and Solutrean points to name just two. Common sense tells us that curated tools are elements of a longer-range technology than noncurated tools and therefore more intellectually taxing. But how far in the future one plans is *not in itself* relevant to the organizational complexity of the task. Short-range plans can be more complex than long-range plans. It is the relation of the elements of the plan, not the length of the forecast, that is crucial. Unless we know more about the long-term strategy of curated-tool-use, their minimum competence remains concrete operational intelligence. The same is true of facilities, stationary technologies like pit falls and fish traps designed to capture without direct human participation (Oswalt 1973). Again, nothing about the construction or geometry requires more than concrete operational intelligence, so we must consider the strategies of use – something not directly observable.

The strategies of use fall under the rubric of subsistence. Here again the Upper Palaeolithic appears more complex than earlier subsistence systems. But does this reflect, as Binford (1982) maintains, a more powerful reasoning ability? Upper Palaeolithic subsistence contrasts with earlier periods in at least two respects: first, some groups (though by no means all) appear to have specialized on gregarious herd animals; and second, towards the end of the Upper Palaeolithic, at least, there is reliance on fishing. Binford (1982) argues that specialization on gregarious mammals like reindeer is linked to a periodic aggregation of the species. At such times they can be exploited more easily, but such mass hunting almost requires some form of storage. Bahn (1977), in a similar vein, argues for selective killing of males in autumn. Such a system must be based on a year-round strategy rather than the short-term hunting or scavenging episodes of previous times. Binford further argues that the emergence of curated technologies in the Upper Palaeolithic corroborates the use of long-range strategies. Fish are difficult to exploit (Dennell 1983), except in spawning seasons with the use of facilities. This is again a matter of long-range plans. These appear to be fair interpretations but the new strategy is not, in fact, more intellectually demanding than hunting episodes of a few days' duration. Piaget studied the development of concepts of time (Piaget 1969a) and concluded that the relations used in constructing a concept of time (it is not perceived!) are the same as those used in constructing space. Of particular importance are spatiotemporal operations, like substitution, similar to those used in conceiving the cross-sections of bifaces (see above). Concrete operations are perfectly capable of constructing temporal frames of years, cycles of seasons, and cycles of game availability. In other words, while common sense may suggest that long-range planning is especially difficult, formal theory argues that the minimum competence is concrete operational intelligence.

The two most easily documented domains of prehistoric behaviour, tech-

nology and subsistence, have provided no evidence for formal operations. I
will now turn to the more elusive domains of social organization and ritual.
Here I will rely entirely upon interpretations, many of which are themselves
controversial. As in the domains of technology and subsistence, one difference
between Upper Palaeolithic social organization and earlier systems is a difference in magnitude, in this case not of time but of space. Bahn (1977) observed
that European Upper Palaeolithic hunters carried or traded shells and other
raw material hundreds of kilometres. The geographic and, presumably social
range of groups was apparently larger than those of earlier periods. Gamble
(1982) argues that the subsistence system of Europe during the Upper Palaeolithic required the exchange of information about far-flung resources and
conditions and that such information could come only from distant kin, real
or fictive. He sees evidence of these regional information-exchange patterns in
the distribution of distinctive artefact styles, which may have operated as
indices of social affiliation (Wynn, in press). Piaget himself rarely commented
on the cognitive prerequisites of social organization. Nevertheless, if we look
at the organizational requirements of exchange systems we must conclude that
the minimum competence was again concrete operations – simple reversibility
in planning and the organization of real, as opposed to hypothetical, information. It is unnecessary for there to have been a general theory of information or style for such a system to work. Simple contingency plans would
suffice.
   It is only in the realm of ritual that we find the glimmerings of organizations beyond the scope of concrete operations. Of course, even simple
interpretations of palaeolithic ritual behaviour are controversial and Leroi-
Gourhan's (1967), the one I choose to examine, is not simple. I am not so
much interested in Leroi-Gourhan's conclusions about dualistic systems and
male–female symbols as I am in his documentation of associations and
repetitions in parietal art. These suggest something interesting about the
prehistoric classification system. For example, 91 per cent of the painted bison
are found in central portions of caves; 64 per cent of the bison are associated
with horses; 'wide signs' dominate the central panels (Leroi-Gourhan 1967,
pp. 112–37). I am aware of the problems of cave topography and Leroi-
Gourhan's occasionally odd method of counting (Ucko & Rosenfeld 1967),
but he does make a good case for certain redundancies in composition, a case
made stronger by its independent discovery by Laming-Emperaire (Leroi-
Gourhan 1967, p. 110). If bison, horses and 'wide signs' do represent a
coherent association of symbolic value, as Leroi-Gourhan maintains, then they
represent a rather sophisticated form of classification. Concrete operational
classification groups members on the basis of tangible similarities, not
hypothetical commonalities. '... even in a zoological classification ... you
cannot extract two noncontiguous classes, like oysters and camels, and make
them into a new "natural" class' (Piaget 1969b, quoted Gruber & Voneche
1977, p. 398). One could argue that horses and bison are 'natural' in this sense,
but this does not appear to be what the Magdalenian painters had in mind.
The animals, the signs and the positions were grouped according to some

abstract common feature (whether or not this is 'femaleness' is irrelevant) and not a tangible similarity. This requires formal operations, at least as Piaget generally defined them. Unfortunately, even if this assessment were true, we have documented formal operational intelligence only for the Magdalenian, perhaps 16 000 years ago, and this is so close to the present as to be unremarkable.

There is evidence of nonutilitarian behaviour prior to the Magdalenian. From the German site of Hohlenstein-Stadel there is a lion-headed anthropomorph carved in ivory that dates to before 30 000 years ago (Marshack 1989), tempting one to posit an iconographic system as subtle as that in Magdalenian art. Marshack makes the persuasive argument that such figures represent ' . . . the end product of a long development'. Engraved bones appear in European sites as early as 190 000 years ago (Gamble 1980). Marshack (1972) has made a case for the notational function of such engraved bones and, leaving aside his interpretation of lunar calendars, it appears that the makers were keeping account of something. Other provocative bits of evidence argue for symbolic behaviour during the Middle Palaeolithic, but interpretations are controversial (Chase & Dible 1987, Marshack 1989). However, none of these demonstrates associational patterns as complex as those of Magdalenian parietal art. If the engraved tablets were notational tallies of some sort, this requires at most a concept of number. Piaget (1952) has shown that a concept of number is based on concepts of class inclusion and seriation, both of which are constructed with the simple reversibilities of concrete operations. Given the tool geometries at 300 000 years ago, the possible appearance of notation at 190 000 is not a surprise. Even lion-headed anthropomorphs are within the abilities of concrete operations (lions and humans share the tangible feature of 'animate being', though I suspect that this is, again, not quite what the Aurignacian carver had in mind). Such evidence suggests a subtle symbolic system, perhaps, but no leap in intelligence.

To summarize, the archaeological evidence for formal operations is mostly negative. None of the technological or subsistence developments that appeared after 300 000 years ago requires more than concrete operational organizations. I include here both developments associated with archaic *H. sapiens* and anatomically modern humans. At most we can conclude that technology and subsistence developed a larger temporal scope. This is not organizational, however, and all remain within the competence of concrete operations. The same is true of the greater geographic scope of certain Upper Palaeolithic social contacts. The only possible evidence for the most abstract organizations of formal operations comes from Magdalenian parietal art. This is based on Leroi-Gourhan's controversial theory and, even if it is sound, it places formal operations so late in prehistory as to be unremarkable. It cannot be used to argue for the intellectual supremacy of the anatomically modern humans over archaic *H. sapiens*.

Because archaeology can document only the minimum necessary competence, it is still possible that anatomically modern humans did, in fact, employ formal operations but that the relevant behaviours have simply left no clues. This is a weak argument in its absence of evidence, but it is at least possible. A closer examination of formal operations, especially the cross-cultural evidence, will, I think, weaken the possibility still further.

Critique of formal operations

Piaget's stage of formal operations is elusive in prehistory, as I have just shown. It is also elusive in the modern world. 'The very few cross-cultural studies that have included tasks of the formal operational stage have found very little evidence of formal operational performance' (Dasen & Herron 1981, p. 332). Concretre operations, on the other hand, appear to be universal in adults. Unfortunately, cross-cultural studies are fraught with methodological problems, especially when using Piagetian tasks. As Piaget observes, few anthropologists are well enough trained to administer the tests and few psychologists are familiar enough with a native people to create a comfortable testing situation (Bringuier 1980). Also, most cross-cultural applications of Piaget are based on a small set of tasks, occasionally only one, and this is insufficient for a reliable assessment (Cole & Scribner 1977). Nevertheless, the absence of formal operations is thought-provoking, especially given the complex behaviours produced by so-called primitive peoples. Micronesian sailors can travel hundreds of miles between tiny atolls using an elaborate system of sidereal navigation, ocean currents, birds, and so on, and yet they do not perform well on Piagetian tasks (Gladwin 1970). Much of the discrepancy appears to be in the domains tested. Balancing a scale is familiar and relevant to Swiss high school students but unfamiliar to a Micronesian sailor. Indeed, not all Western adults perform at the formal operational level (Dasen & Herron 1981). Formal operations may, in fact, be a kind of thinking used by literate, educated adults. Results suggest that some kind of schooling may be necessary for formal operational performance (Rogoff 1981) and, more specifically, that experience in texts may be of crucial significance (Scinto 1984). The nature of texts is self-reflective and forces the writer to transcend content and deal in form. This may in turn lead the individual to use this new style of organization in other domains. Formal operations, then, may be a rather artificial style of thinking one learns in school. This makes it no less useful but does seriously challenge its status as a stage of intellectual development.

The problem we have just encountered is that of disentangling intelligence as some inherent competence from culture as a learned set of solutions, and it bears heavily on our understanding of recent human evolution. Piaget himself was aware of the tangle and indeed incorporated it into his constructivist view of ontogeny. 'Moreover, the history of formal structures [formal operations] is linked to the evolution of culture and collective representations as well as [to] their ontogenetic history' (Inhelder & Piaget 1958, as quoted in Gruber & Vonech 1977, p. 436). Because new cognitive structures are constructed through a dialectic (assimilation and accommodation) between existing structures and the external milieu, some aspect of the milieu must force the disequilibrium in thought that leads to the reorganization of concrete operations into formal operations. Piaget sees the social environment as being crucial in this transition, an environment that has its own history. In Western thought this history includes reliance on texts, formal proofs, deductive logic, and so on. But the history of Micronesian sailors did not include these things and, as a consequence, their final operational structures appear rather different from those of Western adults. This is not to say that the cultural milieu *determines* intelligence, only that it presents rather different problems for intelligence to solve.

So, formal operations may not be universal. They are probably an artefact of operational relations – reversibility, conservation, etc. – applied in a Western educated milieu that has its own history. As a consequence, it may be rather foolish to look for them in prehistory. Nevertheless, the enigma of formal operations does supply some insights into the relationship between intelligence and culture, a relationship that has probably held for some 300 000 years.

Conclusions

Before entering into a more speculative discussion I would first like to review the conclusions of this Piagetian analysis.

The geometry of 300 000-year-old stone tools requires the operational organizations supplied by reversibility and conservation. Preoperational organizations are incapable of conceiving or executing the fine bilateral symmetries and the multiple symmetrical cross-sections of later Acheulean bifaces. The minimum necessary competence was that of concrete operational intelligence. I am not arguing that later Acheulean culture was indistinguishable from modern culture, only that the cognitive organizations behind it were equivalent to those of most modern culture.

None of the Stone-Age developments after 300 000 years ago require an intelligence more sophisticated than concrete operations. I include here technological developments such as prismatic cores, curated tools, and facilities; subsistence developments such as specialization and fishing, both of which require long-term plans; and social developments such as exchange networks. None of these requires more than the mental reversibility and conservation of concrete operations. The only Stone-Age patterns that would have required the more abstract organizations of formal operations are those of Magdalenian parietal art, assuming of course that Leroi-Gourhan's scheme is correct. In other words, from a Piagetian perspective Middle Palaeolithic and Upper Palaeolithic cultures are indistinguishable.

As a consequence, we have no rigorous base from which to argue that anatomically modern humans had some innate capacity for culture that was more powerful than that of their archaic antecedents. Piaget does see a relation between formal operations and physiology. 'It seems clear that the development of formal structures in adolescence is linked to the maturation of cerebral structures. [But] the exact form of linkage is far from simple, since the organization of formal structures must depend on the social milieu as well' (Inhelder & Piaget 1958, quoted in Gruber and Voneche 1977, p. 435). However, the cross-cultural evidence suggests that the social milieu may in fact be the only relevant variable in the acquisition of formal operations. Formal operations are rarely achieved outside of Western educated adults, and are not even universally true for these. It appears to be a style of operational thinking rather than a stage. The basic organizational principles of operational thought appear to be the final stage generally attained by modern humans. The differences between Micronesian sailors and Swiss high school students are a matter of social and cultural milieu, not some inherent difference in the cerebral physiology of the populations. Why should it have been any different in the final periods of the Stone Age?

If physiological evolution cannot be used to account for the documented increase in complexity, what can? Certainly Upper Palaeolithic culture is more complex than Middle Palaeolithic culture in terms of the number of its elements, the temporal range of its subsistence, and the geographic range of its social contacts. But so is Western culture compared with that of highland New Guinea and so is 20th-century technology compared with that of the 19th. It is culture itself that has become more complex and, odd as it sounds, more intelligent. The complex social and technological fabric of which we are a part allows us to solve complex organizational problems: 'the most generalized forms of thought, those that can be dissociated from their content, are, by that very fact, forms of cognitive exchange or of interindividual regulation' (Piaget 1971, pp. 360–1). This 'interindividual regulation' has, I maintain, been the crucial component of human behaviour for the last 300 000 years. It has increased in scope and organizational power, but this is not a matter of brain evolution. We cannot disentangle the evolution of intelligence from the evolution of culture because, from 300 000 years ago, they are one and the same thing. Upper Palaeolithic culture may have been more finely adapted than that of the Middle Palaeolithic, but this was not because the participants were more intelligent. Their culture was simply different and, it appears, in the long run more successful.

References

Bahn, P. G. 1977. Seasonal migration in southwest France during the late glacial period. *Journal of Archaeological Science* 4, 245–57.

Binford, L. R. 1982. Comment on 'Rethinking the Middle Upper Palaeolithic transition' by R. White. *Current Anthropology* 23, 177–81.

Binford, L. R. 1983. *In pursuit of the past*. London: Thames & Hudson.

Bringuier, J. 1980. *Conversations with Jean Piaget*. Chicago: University of Chicago Press.

Chase, P. & H. Dibble 1987. Middle Palaeolithic symbolism: a review of current evidence and interpretations. *Journal of Anthropological Archaeology* 6, 263–96.

Cole, M. & S. Scribner 1977. Developmental theories applied to cross-cultural research. *New York Academy of Sciences Annals* 285, 366–73.

Dasen, P. 1977. *Piagetian psychology: cross-cultural contributions*. New York: Garden Press.

Dasen, P. & A. Herron 1981. Cross-cultural tests of Piaget's theory. In *Handbook of cross-cultural psychology: vol. 4 developmental psychology*, H. Triandis & A. Herron (eds), 295–341. Boston: Allyn & Bacon.

Dennell, R. 1983. *European economic prehistory*. London: Academic Press.

Foley, R. A. 1991. How useful is the culture concept in early hominid studies? In *The origins of human behaviour*, R. A. Foley (ed.), Ch. 2. London: Unwin Hyman.

Gamble, C. 1980. Information exchange in the palaeolithic. *Nature* **283**, 522–3.

Gamble, C. 1982. Interaction and alliance in palaeolithic society. *Man* **17**, 92–107.

Gladwin, T. 1970. *East is a big bird*. Cambridge, Ma.: Harvard University Press.

Gould, S. 1977. *Ontogeny and phylogeny*. Cambridge, Ma.: Harvard University Press.

Gould, S. 1981. *The mismeasure of man*. New York: Norton.

Gruber, H. & J. Voneche 1977. *The essential Piaget*. New York: Basic Books.

Inhelder, B. & J. Piaget 1958. *The growth of logical thinking from childhood to adolescence*. (Trans. A. Parsons & S. Milgram.) New York: Basic Books.

Leroi-Gourhan, A. 1967. *Treasures of prehistoric art*. (Trans. N. Guterman.) New York: Abrams.

Marschack, A. 1982. Upper Palaeolithic notation and symbol. *Science* **178**, 817–28.

Marshack, A. 1989. Evolution of the human capacity: the symbolic evidence. *Yearbook of Physical Anthropology* **32**, 1–34.

Oswalt, W. 1973. *Habitat and technology*. New York: Holt, Rinehart & Winston.

Piaget, J. 1952. *The child's conception of number*. (Trans. C. Gattegno & F. Hodgson.) London: Routledge & Kegan Paul.

Piaget, P. 1969a. *The child's conception of time*. (Trans. A. Pomerans.) London: Routledge & Kegan Paul.

Piaget, J. 1969b. *The psychology of the child*. (Trans. H. Weaver.) London: Routledge & Kegan Paul.

Piaget, J. 1970. *Structuralism*. (Trans. C. Maschler.) New York: Harper.

Piaget, J. 1971. *Biology and knowledge*. Chicago: University of Chicago Press.

Piaget, J. 1972. *The principles of genetic epistemology*. (Trans. W. Mays.) London: Kegan Paul.

Piaget, J. & B. Inhelder 1967. *The child's conception of space*. (Trans. F. Langlon & J. Lunzer.) New York: Norton.

Redman, C. 1978. *The rise of civilization*. San Francisco: Freeman.

Rogoff, B. 1981. Schooling and the development of cognitive skills. In *Handbook of cross-cultural psychology: vol. 4 developmental psychology*, H. Triandis & A. Herron (eds), 233–94. Boston: Allyn & Bacon.

Scinto, L. 1984. The architectonics of texts produced by children and the development of higher cognitive functions. *Discourse Processes* **7**, 371–418.

Ucko, P. & A. Rosenfeld 1967. *Palaeolithic cave art*. New York: McGraw Hill.

Wynn, T. 1979. The intelligence of later Acheulean hominids. *Man* **14**, 371–91.

Wynn, T. 1981. The intelligence of Oldowan hominids. *Journal of Human Evolution* **10**, 529–41.

Wynn, T. 1985. Piaget, stone tools, and the evolution of human intelligence. *World Archaeology* **17**, 31–43.

Wynn, T. 1989. *The evolution of spatial competence*. Urbana: University of Illinois Press.

Wynn, T. In press. The evolution of tools and symbolic behaviour. In *The evolution of human symbolic behaviour*, A. Locke & C. Peters (eds). Oxford: Oxford University Press.

5 The invention of computationally plausible knowledge systems in the Upper Palaeolithic

SHELDON KLEIN

In recent years, with the expansion of computer science, researchers in the cognitive sciences have been attracted towards the use of computational models for understanding the structure of human thought. Such work as has been done has proved extremely powerful for tackling contemporary situations (Haugeland 1985), and so it is appropriate to ask whether such approaches have the potential for explaining the evolutionary development of systems of human knowledge. In this chapter I shall examine how such knowledge systems may be structured, and whether there is evidence for their origins in human prehistory.

The problem of computing human behaviour by rules

Contemporary artificial-intelligence researchers find the problem of computing human behaviour by rules intractable for large-scale knowledge systems. While excellent results have been obtained for small-scale knowledge domains, the time it takes to make such computations can increase exponentially or even combinatorially with the size and heterogeneity of the knowledge system. If the human brain, like a computer, is a finite-state automaton, then the problem of generating and parsing behaviour must present the same computational difficulty for the human mind.[1] The problem of making such computations at a pace fast enough for ordinary social interaction can be solved if appropriate constraints apply to the structure of the rules. There seems to be evidence that systems of such constraints were invented in the Upper Palaeolithic, and were of such power as to guarantee that the time necessary for computation of behaviour would increase only linearly with the size and heterogeneity of the world knowledge systems. The evidence can be found in the material and symbolic artefacts of a variety of cultures, and the major sources are classification schemes, divination systems, iconographic systems, language structures, and shamanistic, mythic, or religous systems.

The purpose of this chapter is to establish a model by which the complexities of human behaviour can be generated using a system of rules that is consistent with how human thought operates; is parsimonious, allowing for the processing and manipulation of knowledge to occur rapidly; is internally consistent; and permits knowledge to be accumulated. In other words, this is an attempt to construct a model of the mind that is capable both of being practical in computational terms and of accounting for the heterogeneity in human knowledge systems. The key attribute for this lies in the use of rules governing the association and transformation of items of knowledge. This in turn rests on the use of formal logic for treating the classification of knowledge, and consequently such formal logic provides the methodological framework for what follows.

The basic structure of the invention

Fundamentally, there was one computational invention, capable of unifying the full range of human sensory domains and consisting of an analogical reasoning method used in combination with global classification schemes. The structure of the human brain may be a factor in the history of this invention, but its utility exists independently of such a connection. Every culture seems to have a global classification scheme in the history of its knowledge structures and usually such schemes can be linked to myth systems. The use of this invention to compute human behaviour is explained fully elsewhere (Klein 1983, 1988). The strong equivalence operator of logic is shown to define ATOs (appositional transformation operators) that relate the input and output states of behavioural rules by analogical transformations. It is argued that a given culture has a relatively small set of such ATOs and that they apply to diverse domains of human behaviour, with a processing time that increases only linearly with the number of elements relevant to those rules. The global classification scheme makes it possible to select and apply the appropriate ATOs in a variety of domains by specifying equivalence classes of elements that may serve as substitution sets for the extension of each ATO. The result can be compared to a set of canonical analogies for which the extension and application are determined by equivalent analogues in the global classification scheme. The classification scheme for Chinese culture (Table 5.1) is a typical example (Klein, 1983, p. 159).

Each semantic domain is seen to have its equivalent in another domain. For example, 'East' is the *direction* counterpart of the *element* 'wood', and its *season* companion is 'spring'. Each of these terms is itself a metonym representing another class of items. The Chinese scheme is also linked to the I Ching divination system, which may be viewed as a knowledge-based query system based on analogical principles. The divination system is associated with a set of canonical texts containing specific terms of reference that function as metonyms for higher-level classes. Each text may be viewed as a formulaic behaviour pattern awaiting the substitution of appropriate values for its variable terms by the user of the divination system. The computationally difficult problem is the selection of a culturally consistent set of elements for

Table 5.1 Some trigram correspondences.

	001 thunder	110 wind	101 fire	100 mountain	000 earth	111 heaven	011 lake	010 water
Element	wood		fire	earth			metal	water
Direction	east		south	centre			west	north
Colour	blue		red	yellow			white	black
Season	spring		summer	'jang'			autumn	winter
Climate	windy		hot	humid			dry	cold
Planet	Jupiter		Mars	Saturn			Venus	Mercury
Sound	shouting		laughing	singing			weeping	groaning
Musical note	*chüeh*		*chih*	*kung*			*shang*	*yü*
Emotion	anger		joy	sympathy			grief	fear
Animal	dragon	fowl	pheasant	dog	ox	horse	sheep	pig
Family	1st son	1st daughter	2nd daughter	3rd son	mother	father	3rd daughter	2nd son
Body part	foot	thigh	eye	hand	belly	head	mouth	ear
Attribute	movement	penetration	brightness	standstill	docility	strength	pleasure	danger

Sources: Blofeld 1978, pp. 190–1, Legge 1964 [1899], pp. xliv–v, Legeza 1975, p. 11, Fung Yu-lan 1953 [1934], pp. 40–2, 86–132.

the terms in the text. For a computer program operating with rules formula-ted in propositional logic, this could involve a combinatoric computation process. The Chinese global classification scheme reduces the process to looking up the corresponding elements in a table. However, the classification scheme used in a given divination is actually a transformation of the basic one shown in Table 5.1. The divination process yields an ATO which generates an analogical realignment of the original table in correspondence with the situation of the moment, as determined by the divination process. A wide-spread African divination system operates on the same principles, and they can be seen to work also in the visual and verbal iconography of Navaho curative ceremonies. Tibetan and esoteric Japanese Buddhist iconography functions as an ATO system which is visual encoding of ATOs applicable to specific world domains, in conjunction with a myth system and a global classification scheme (Klein 1983).

At this point let me offer some intuitive examples of how ATOs work in verbal and visual analogical reasoning problems, and also examples of analogi-cal computation of behaviour using situation descriptions linked by ATOs (Klein, 1983, pp. 152–4).

ATOs relate situation descriptions in the form of arrays of features. A two-valued version can be defined by the strong-equivalence operator of logic, which can be used to compute ATOs:

$$\star \quad A \quad B \; = \; C$$

0	0	1
0	1	0
1	0	0
1	1	1
.	.	.

The '$\star$' means that a result is to be computed using the above truth table. A '1' means 'true', a '0' means 'false', and '.' means 'does not apply'. ATOs may also be computed with the rules for binary addition (mod-2 arithmetic), if the interpretations of 1 and 0 are reversed.

The operator is commutative, $\star AB \; = \; \star BA$
involutive, $\star A(\star AB) \; = \; B$
and associative, $\star A(\star BC) \; = \; \star(\star AB)C$

and has a mathematical group property. Consider, for example, the ATO relating two hypothetical feature arrays, A and B. Each feature value in A is matched with its positional counterpart in B to compute its component in the ATO $\star AB$:

$$
\begin{array}{cc}
A & B \\
110 & 011 \\
011 \xleftrightarrow{} & 100 \\
1.1 & 0.0
\end{array}
$$

$$
\begin{array}{c}
\star AB \\
010 \\
000 = \text{ATO} \\
0.0
\end{array}
$$

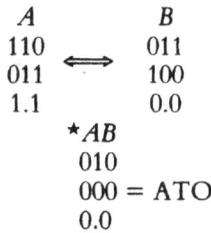

Some simple analogies will illustrate how ATOs work (Klein 1983, pp. 152–4):

A feature array referencing 'male', 'female', 'young', 'adult', 'love', 'hate', 'light', and 'dark' is sufficient to formulate the following analogy:

$$
\frac{X = \text{boy loves light}}{Y = \text{woman hates light}} \; :: \; \frac{Z = \text{girl hates dark}}{?}
$$

M	F	Y	A	L	H	Lt	Dk	where M = male, F = female,

Y = young, A = adult, L = love,
H = hate, Lt = light, Dk = dark

$$
\frac{X = 10101010}{Y = 01010110} \; :: \; \frac{Z = 01100101}{?} \star XY = 00000011
$$

$$
? = \star Z(\star XY) = 10011001 = \text{man loves dark}
$$

Another example:

$$
\frac{X = \text{man hates dark}}{Y = \text{woman loves light}} \; :: \; \frac{Z = \text{boy hates light}}{?}
$$

$$
\frac{X = 10010101}{Y = 01011010} \; :: \; \frac{Z = 10100110}{?} \star XY = 00110000
$$

$$
? = \star Z(\star XY) = 01101001 = \text{girl loves dark}
$$

The same method can be applied to visual analogies. For example, if a set of visual features is used to create a pictorial analogy (Fig. 5.1), the answer can be calculated using ATOs (Fig. 5.2). If we give natural-language interpretations to these visual features, we can obtain the results shown in Figure 5.3.

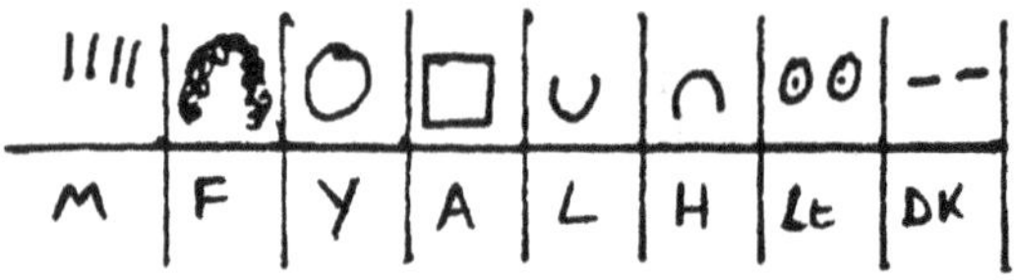

Figure 5.1 A pictorial analogy

Key: M = male; F = female; Y = young; A = adult; L = love; H = hate; Lt = light; Dk = dark.

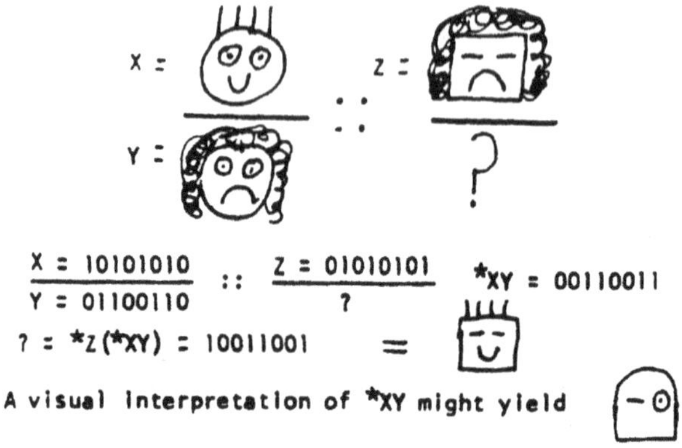

$$\frac{X = 10101010}{Y = 01100110} :: \frac{Z = 01010101}{?} \quad {}^*XY = 00110011$$

$$? = {}^*Z({}^*XY) = 10011001 \quad =$$

A visual interpretation of *XY might yield

Figure 5.2 Calculation of a pictorial analogy.

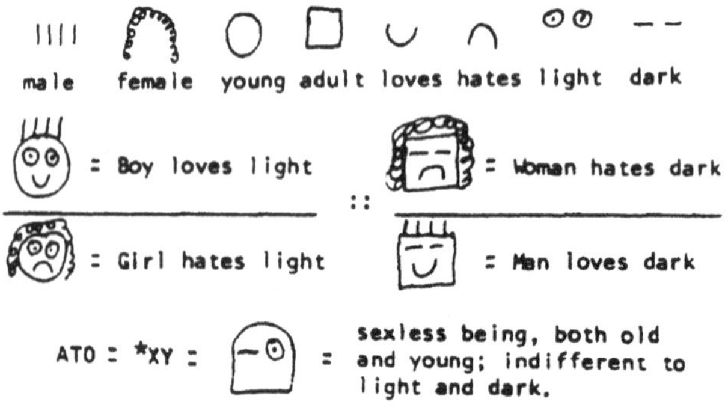

Figure 5.3 The pictorial analogy with a natural-language interpretation.

Complex analogies may also be computed, as in the following abstract example:

$$\text{If } ((X :: Y) :: (Z :: W)) :: (P :: ?), \text{ then}$$
$$? = {}^*P({}^*(XY)({}^*ZW))$$

A concrete illustration of this abstract example is as follows:

	X	Y
	A loves B, has no $, and is unmarried. B loves A, has no $, and is unmarried. C loves no one, has $, and is unmarried.	A loves B, has no $, is married to B. B loves A, has no $, is married to A. C loves no one, has $, and is unmarried.

Where La means 'loves A', etc., $ means 'has money', and Ma means 'married to A', etc., the X and Y states may be represented as follows:

	La	Lb	Lc	$	Ma	Mb	Mc
A	.	1	0	0	.	0	0
B	1	.	0	0	0	.	0
C	0	0	.	1	0	0	.

X

$\Rightarrow$

	La	Lb	Lc	$	Ma	Mb	Mc
A	.	1	0	0	.	1	0
B	1	.	0	0	1	.	0
C	0	0	.	1	0	0	.

Y

$\star XY$

.	1	1	1	.	0	1
1	.	1	1	0	.	1
1	1	.	1	1	1	.

If we depict 'loves' as a nose pointing at the beloved (in between, if two loves), if a noseless state means 'loves no one', if holding hands depicts 'married to', and if a '$' indicates 'has money', we obtain the visual interpretation of Figure 5.4.

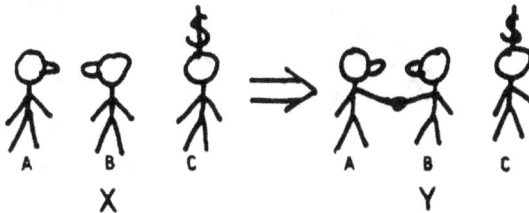

Figure 5.4 A visual interpretation of $X \rightarrow Y$, where X is 'A loves B, has no $, and is unmarried. B loves A, has no $, and is unmarried. C loves no one, has $, and is unmarried' and Y is 'A loves B, has no $, and is married to B. B loves A, has no $ and is married to A. C loves no one, has $, and is unmarried'.

Continuing with this complex example:

Z		W
A loves no one, has no $ and is married to *B*. *B* loves *A*, has no $, and is married to *A*. *C* loves *A*, has $, and is unmarried.	$\Rightarrow$	*A* loves no one, has $, and is married to *C*. *B* loves no one, has no $, and is unmarried. *C* loves *A*, has $, and is married to *A*.

	La	Lb	Lc	$	Ma	Mb	Mc
A	.	0	0	0	.	1	0
B	1	.	0	0	1	.	0
C	1	0	.	1	0	0	.

Z

$\Rightarrow$

	La	Lb	Lc	$	Ma	Mb	Mc
A	.	0	0	1	.	0	1
B	0	.	0	0	0	.	0
C	1	0	.	1	1	0	.

W

$\star ZW$

.	1	1	0	.	0	0
0	.	1	1	0	.	1
1	1	.	1	0	1	.

This yields the visual interpretation of Figure 5.5.

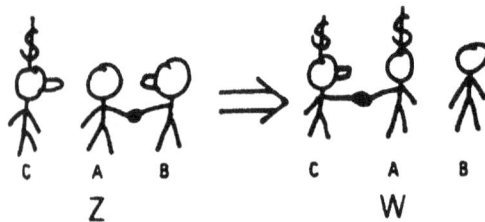

Figure 5.5 A visual interpretation of $Z \rightarrow W$, where Z is '*A* loves no one, has no $, and is married to *B*. *B* loves *A*, has no $, and is married to *A*. *C* loves *A*, has $, and is unmarried' and W is '*A* loves no one, has $, and is married to *C*. *B* loves no one, has no $ and is unmarried. *C* loves *A*, has $, and is married to *A*.'

$^{*}(^{*}XY)(^{*}ZW)$ 'surrealistic' interpretation

$$\frac{\begin{array}{c} .\ 1\ 1\ 0\ .\ 1\ 0 \\ \hline 0\ .\ 1\ 1\ 1\ .\ 1 \end{array}}{1\ 1\ .\ 1\ 0\ 1\ .} = \begin{array}{l} A \text{ loves } B \text{ and } C, \text{ has no \$, and is married to } B.\ B \\ \text{loves } C, \text{ has \$, and is married to } A \text{ and } C.\ C \text{ loves } A \\ \text{and } B, \text{ has \$, and is married to } B. \end{array}$$

The visual interpretation obtained is that in Figure 5.6

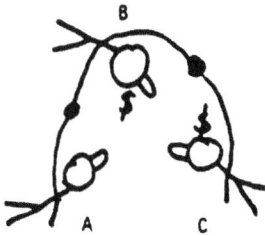

Figure 5.6 A visual interpretation of the 'surrealistic' interpretation' $^{*}(^{*}XY)\ (^{*}ZW)$: 'A loves B and C, has no \$, and is married to B. B loves C, has \$, and is married to A and C. C loves A and B, has \$ and is married to B'.

If we then postulate a situation P,

	La	Lb	Lc	\$	Ma	Mb	Mc	
A	.	1	1	0	.	0	0	
B	1	.	0	0	0	.	0	=
C	1	0	.	1	0	0	.	

A loves B and C, has no \$, and is unmarried. B loves A, has no \$, and is unmarried. C loves A, has \$, and is unmarried.

we can compute its successor state by analogy with the combined results of X—Y and Z—W by solving

$$((X :: Y) :: (Z :: W)) :: (P :: ?)$$

where $? = {}^{*}P(^{*}(^{*}XY)(^{*}ZW))$, which can be represented as follows:

	La	Lb	Lc	\$	Ma	Mb	Mc	
A	.	1	1	1	.	0	1	
B	0	.	0	0	0	.	0	=
C	1	0	.	1	1	0	.	

A loves B and C, has \$, and is married to C. B loves no one, has no \$, and is unmarried. C loves A, has \$, and is married to A.

This yields Figure 5.7.

Figure 5.7 A visual interpretation of $P \rightarrow {*}P({*}({*}XY)({*}ZW))$, where P is 'A loves B and C, has no \$, and is unmarried. B loves A, has no \$, and is unmarried. C loves A, has \$, and is unmarried' and ${*}P({*}({*}XY)({*}ZW))$ is 'A loves B and C, has \$, and is married to C. B loves no one, has no \$, and is unmarried. C loves A, has \$, and is married to A'.

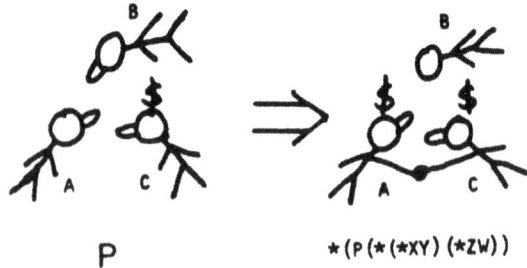

$$P \qquad {*}(P({*}({*}XY)({*}ZW)))$$

ATOs, language, and culture

I wish to argue that the invention of computational knowledge, consisting of the idea of a global classification scheme, in combination with behaviour rules related by a limited set of analogical transformation operators, was responsible for the elaboration of language and culture structures in a process of coevolution. Phrase-structure grammar operates on ATO principles; this can be verified by creating a categorial grammar in which grammar codes consist of appropriately chosen binary integers. If one adds information indicating right- or left-combining properties and also adds semantic-feature vectors, it is possible to use ATO logic for decoding both syntax and semantics in the same notation. An implication is that world knowledge systems and language systems have coevolved. If this is so, then:

1   The Sapir–Whorf hypothesis that the structure of grammar determines world view may remain true synchronically; diachronically, however, the two systems are in an intimate relationship of mutual influence and modification.
2   While the ATO model does not 'refute' Chomsky's view that there is an innate, genetic basis for language structure, it makes that assumption unnecessary to account for human linguistic behaviour. The structure of the human brain may be a passive factor in the invention of structures that are computationally efficient in a given 'hardware' environment.

The extension and elaboration of culture content can be interpreted as the extension of the global classification scheme to new elements, and as the application of existing ATO patterns to new behavioural situations. The result is a formally definable explanation of the process of creating new patterns of behaviour by analogy with patterns in other domains. If this process is part of the growth of a culture and its social institutions, then its symbolic, behavioural, and material artefacts will contain many homologies. It is this aspect that gives a culture its coherency and enables its members to know what culture elements are appropriate.

ATOs and the ontogeny of shamanism

Religious systems can be interpreted as the symbolic medium in which ATO systems are encoded. The hierarchy of ATOs that govern the structure of a culture are inevitably encoded surrealistically in verbal and plastic domains, including myth systems and representations of spirits and deities.

Consider the following aspects of the computation of behaviour with ATOs (Klein, 1983, p. 154).

If a sequence of events, A, B, C, D, occurs, then:

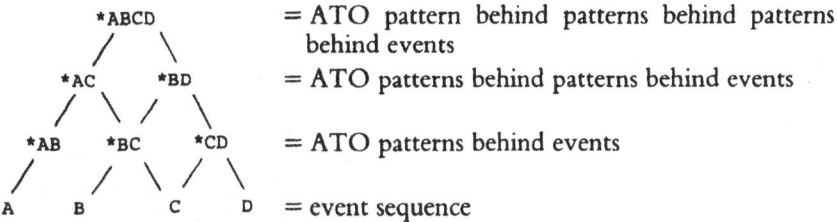

```
     *ABCD              = ATO pattern behind patterns behind patterns
    /    \                 behind events
 *AC      *BD           = ATO patterns behind patterns behind events
 / \    / \
*AB  *BC   *CD          = ATO patterns behind events
/   / \  / \
A   B   C   D           = event sequence
```

If we wish to obtain a state E instead of D, *without changing* any of the ATOs, we derive, by analogy, a sequence leading to E by replacing A, B, C respectively with $^*A(^*DE)$, $^*B(^*DE)$, $^*C(^*DE)$. If we wish to make a plan that specifies more than one goal state in the event sequence, we must alter some ATOs.

The meaning of 'culturally defined behaviour', is that members of a society plan in a way that minimizes the level and number of ATOs affected. It follows that deviant behaviour may be interpreted as behaviour that violates acceptable levels and numbers of ATOs. ATO patterns are part of the knowledge acquired by children. They are encoded in multiple media of expression, both material and symbolic, and are the source of metaphor. It is this encoding that gives form to a culture, and it is the widely distributed presence of ATOs in the environment that makes calculation of social behaviour computationally feasible for the human mind.

The emergence of a canonical hierarchy of ATOs, applicable to multiple domains of social reality through the mediation of a global classification scheme, would be a natural consequence of organizing social life on the basis of ATO logic. If we make the assumption that the human mind encodes ATOs in iconic imagery, we may also suggest that such imagery is given metaphysical interpretation. A hierarchic ATO system may be interpreted by the human mind as hierarchy of spiritual beings, and the spirit journey of a shaman seeking to resolve problems in a spirit realm can be interpreted as precisely the kind of ATO manipulation described above. Magic spells and rituals would appear as devices for inserting desirable ATOs in given situations, and it might be possible to predict their form and general content from the global classification scheme. The implication of this model is that shamanism is a consequence of the adoption of computationally plausible knowledge systems. Several theoretical possibilities are implied:

(1)   The ATO system concept was invented once, and spread by diffusion.
(2)   Computation with ATO logic may be a part of the functioning of the human brain.
(3)   ATO systems may have been invented independently, in conjunction with elaboration of social life.
(4)   If (2) and (3) are true, then the concept of 'the shamanistic tradition' may reflect phenomena which are of independent origin (Eliade 1964, *Artscanada*, 1973/4).

The evidence of Lévi-Strauss

The ATO logic I have described in more detail elsewhere (Klein 1983) is a model of the structuralism of Claude Lévi-Strauss. It was originally formulated in 1976–7 in an attempt to replicate the reasoning processes that Lévi-Strauss used in *Mythologiques* (Lévi-Strauss 1964–71). Given his semantic units, the arguments linking myth structures can be verified and replicated by ATO computation (Klein 1977). My 1983 paper was intended as a validation of the ATO concept with independent data. *La pensée sauvage* (Lévi-Strauss 1962) is an explication of human reasoning with ATO systems; the four volumes of *Mythologiques* represent an overwhelming body of empirical evidence that ATO systems exist. The work is an analogue of historial reconstruction linguistics. While he does not reconstruct a protosystem, Lévi-Strauss has demonstrated that proto-ATO systems must have existed at least as early as the Upper Palaeolithic and that they have contemporary descendants. Given this perspective, much work seemingly critical of the structuralism of Lévi-Strauss can be reinterpreted as supportive (Hodder 1982, Miller 1982, Tilley 1982, Wylie 1982).[2]

Testing the ATO model in historical time

My discussion elsewhere of ways one might obtain empirical validation of the ATO concept (Klein 1983, p. 178) includes the following observations:

(4)   the ATO model can be used as a heuristic device to suggest cultural correlations that can be verified by other methods. This approach might even extend to predictions about the location of buildings with specific functions in archaeological sites. Analysis of symbolic artefacts by ATO logic might help to decode or unlock large systems of correlations . . .

(5)   One might examine the possibility that ATOs can be sources of *social and cultural change*. A large-scale classification system can imply a structured universe which no participant in a culture can contemplate as a whole. If a classification system incorporates 50 features, it can imply a conceptual universe with $2^n = 2^{50}$ elements. ATOs that function in a

subset of the implied universe can be used as an exploratory tool to extend knowledge by analogy. A sudden, externally caused change in iconography (or mythology) would imply a new system of correlations and would offer the potential for new analogies about the structure of the world that might imply new patterns of behaviour. A test of such a possibility would require an adequately documented historical situation.

Major testing of the theory requires a detailed analytic perusal of broad streams of history in a number of cultures. I would cautiously cite Toynbee (1934–61) and Spengler (1926–8), whose general theoretical analyses can be interpreted in an ATO framework. I do not endorse any particular details of their analyses, but rather note that in their surveys of massive amounts of data they found relations and structures which are compatible with the theory of ATO systems. The principle that I value in Toynbee is his relation of religious systems to sociocultural systems (after disassociating his ideas from his personal religious bias). In the case of Spengler I value the perception of the analogical relationships among the artefacts of a culture (Spengler 1926, p. 47):

From this moment on, relations and connexions – previously often suspected, sometimes touched on but never comprehended – presented themselves in ever-increasing volume. The forms of the arts linked themselves to the forms of war and state-policy. Deep relations were revealed between political and mathematical aspects of the same culture, between religious and technical conceptions, between mathematics, music, and sculpture, between economics and cognition-forms. Clearly and unmistakably there appeared the fundamental dependence of the most modern physical and chemical theories on the mythological concepts of our Germanic ancestors, the style-congruence of tragedy and power-technics and up-to-date finance, and the fact (bizarre at first but soon self-evident) that oil-painting perspective, printing, the credit system, long-range weapons, and contrapuntal music in one case, and the nude statue, the city-state, and coin-currency (discovered by the Greeks) in another were identical expressions of one and the same spiritual principle.

Conclusions

The criterion that a model of human cognition must account for the ability of humans to compute social behaviour in real time has, to my knowledge, not been addressed before. The thesis that ATO systems were invented in the Upper Palaeolithic and are responsible for the growth of sociocultural structures provides a mechanism for a variety of seemingly disparate theories. It makes structuralism and systems anthropology appear as different aspects of the same phenomenon and, if ATO logic proves to be hardwired in the human brain, it will be particularly compatible with sociobiology.[3]

Notes

1   An assumption that the brain is a massively parallel computer does not mitigate the problem. The addition of n parallel processors can reduce the computation time by a factor of n, but the problem domain involves a processing time that can increase combinatorially with the size of the data base. If an additional computer processor is added for *each* new item in the data base, the processing time may increase at a rate of $n!/n = (n - 1)!$ A connectionist brain model presents an analogous difficulty: the need for combinatorially increasing processing time is replaced by a need for combinatorially increasing connectivity.

2   The seemingly supportive evidence of Leroi-Gourhan (1965) is not supportive because it is not substantiated by knowledge of the global classification scheme of the culture that produced the Lascaux paintings. A recent discussion of the evidence is contained in Marshack (1985, pp. 538–9).

3   A very recent analysis suggests that a developmental sequence in lithic technology dating to the Middle/Upper Palaeolithic transition in the Negev reflects the group concept and ATOs in the cognitive processes of the concerned tool-makers (Klein 1990).

References

Artscanada 1973/4. *Stones, bones & skin: ritual and shamanic art,* **184–7**. 30th anniversary issue.

Blofeld, J. 1978. *Taoism: the road to immortality.* Boulder: Shambhala.

Eliade, M. 1964. *Shamanism: archaic techniques of ecstasy.* Princeton: Princeton University Press.

Haugeland, J. 1985. *Artificial intelligence: the very idea.* Cambridge, Ma.: MIT Press.

Hodder, I. 1982. Theoretical archaeology: a reactionary view. In I. Hodder (ed.), *Symbolic and structural archaeology,* 1–16. Cambridge: Cambridge University Press.

Klein, S. 1977. Whorf transforms and a computer model for propositional/ appositional reasoning. Paper presented at the Applied Mathematics Colloquium, University of Bielefeld, at the Computer Science Colloquium, University of Paris–Orsay, and at a joint colloquium of the Anthropology and Computer Science Department, University of California, Irvine.

Klein, S. 1983. Analogy and mysticism and the structure of culture. *Current Anthropology* **24**, 151–80.

Klein, S. 1988. Reply to S. D. Siemens' critique of S. Klein's 'Analogy and mysticism and the structure of culture', *Current Anthropology* **29**, 478–83.

Klein, S. 1990. Human cognitive changes at the Middle/Upper Palaeolithic transition: the evidence of Boker Tachtit. In *The emergence of modern humans: the archaeological perspective,* P. A. Mellars (ed.), 499–516. Edinburgh: Edinburgh University Press.

Lévi-Strauss, C. 1962. *La pénsee sauvage.* Paris: Plon.

Lévi-Strauss, C. 1964–71. *Mythologiques.* 4 vols. Paris: Plon.

Legeza, L. 1975. *Tao magic: the Chinese art of the occult.* New York: Pantheon Books.

Legge, J. (trans.) 1964 (1899). *The Yi King,* 2nd edn. New Hyde Park, New York: University Books.

Leroi-Gourhan, A. 1965. *Préhistoire de l'art occidental.* Paris: Mazenod.

Marshack, A. 1985. More on serpents in the mind. *Current Anthropology* **26**, 537–9.

Miller, D. 1982. Artefacts as products of human categorization processes. In *Symbolic and structural archaeology*, I. Hodder (ed.), 17–25. Cambridge: Cambridge University Press.

Spengler, O. 1926–8 (1918–22). *The decline of the West*. Vol. 1. 1926(1918). *Form and actuality*. Vol. 2. 1928(1922). *Perspectives of world-history*. New York: Alfred A. Knopf.

Tilley, C. 1982. Social formation, social structures and social change. In *Symbolic and structural archaeology*, I. Hodder (ed.), 26–38. Cambridge: Cambridge University Press.

Toynbee, A. J. 1934–61. *A study of history*. 12 volumes. London: Oxford University Press.

Wylie, M. A. 1982. Epistemological issues raised by a structuralist archaeology. In *Symbolic and structural archaeology*, I. Hodder (ed.), 39–46. Cambridge: Cambridge University Press.

Yu-lan, Fung 1953 (1934). *A history of Chinese philosophy*. Vol. 2. (Trans. Derk Bodde.) Princeton: Princeton University Press.

6 An interactive growth model applied to the expansion of Upper Palaeolithic populations

EZRA B. W. ZUBROW

Outlined against a blue-gray October sky, the Four Horsemen rode again ... In dramatic lore they are known as *Famine, Pestilence, Destruction, and Death*. These are only aliases.

Grantland Rice

And power was given unto them over the fourth part of the earth, to kill with sword, and with hunger, and with death, and with the beasts of the earth.

Revelation 6.8

There has been considerable speculation on the relationship of the two subspecies *Homo sapiens sapiens* and *Homo sapiens neanderthalis* during the period just prior to the Neanderthals' extinction. There is little fact. This chapter uses a simulation model to create possible scenarios for the interaction of the two species at different locations in Europe at about 30 000 BC. The models indicate that there is a very small window which existed in the growth and interaction rates of the two species which would have allowed the Neanderthals to continue. Furthermore, this chapter suggests that the Neanderthal demise was more likely the result of small numbers and chance in a competitive situation than lack of adaptive characteristics. Finally, it suggests that one advantage that *H. sapiens sapiens* had was its more rapid rate of attaining demographic and geographic stability. As is the case in all simulations, reality is modelled; it is not re-created. Therefore, this chapter admittedly contributes to the realm of speculation rather than that of fact.

The background

Since 1856 scholars have been aware of the unusual skeleton found at a quarry in the Neander valley near Düsseldorf. Now with more than 100 sites analysed, a broadly drawn picture of Neanderthal adaptation has been developed through the efforts of numerous archaeologists and physical anthropologists. A brief sketch would note that they were hunter-gatherers with the emphasis probably on gathering, lived in small family bands, made stone tools with Mousterian techniques, and were sufficiently sophisticated to bury their dead. It was a successful adaptation surviving major changes in climate. Neanderthals appeared in Europe about 125 000 years ago and became extinct approximately 30–35 000 years ago. During their 100 000-year existence there was a sufficient geographical radiation for them to have been found in Europe, the Middle East, and Asia.

Early *H. sapiens sapiens* such as Cro-Magnon generally correspond to the Upper Palaeolithic in Europe. From approximately 35 000 to 10 000 years BP, cultural variation increased, as indicated by the diversity of the Perigordian, Aurignacian, Magdalenian, and Solutrean cultures as well as by the increasingly functionally specific types of sites. As hunters and gatherers they were able to adapt to both the climate of the last glaciation in Europe and the warming which followed. It has been assumed that they lived in small bands of about 75 to 100. Ethnographic analogy has suggested that labour use was relatively efficient and that their existence was not limited to a Malthusian minimum. To some extent this viewpoint is substantiated by the great art of the period at places such as Lascaux and by the rapid adaptive, and eventually cultural diversity, of our species.

Considerable interest has been expressed in the transitional examples of hominids. It has long been suggested that the Neanderthals found at Tabun and Amud in Israel were aberrant. They are more similar to *H. sapiens sapiens* than are many other skeletons. Similarly, the sites of Skhul and Qafzeh contain classic Neanderthal Mousterian tools but modern hominids. Alternatively, at St Césaire one finds Neanderthal skeletons associated not with Mousterian but with Chatelperronean stone tools.

The stage is thus set to enter the realm of speculation and consider what the relationships between the two populations may have been. There are several possibilities, which include:

1   *H. sapiens sapiens* and Neanderthals are two distinct populations with the former deriving from the ancestral latter.
2   *H. sapiens sapiens* and Neanderthals are two distinct and partially contemporaneous populations in which the latter became extinct due to competitive pressures from the former.
3   *H. sapiens sapiens* and Neanderthals are one ancestral population and the *sapiens sapiens* characteristics survived due to adaptive or competitive advantage.

The rest of this chapter will be concerned with examining how the simulation model addresses this transition and these three possibilities.

The model

Imagine two bands of hominids moving through a Pleistocene landscape following their respective game animals. The sun rises and falls on their respective camps. As the seasons pass, each traverses a route through their territory. These routes are established by the schedule of harvesting wild plants, game routes, predators, and the location of water. They are also determined by a variety of imponderables: volition, religion, idiosyncratic personality, and simple chance.

These populations are not static. They grow and decline; they break up and reaggregate. This depends on many factors: the local environment, the skill of the subsistence gatherers, disease, and the fertility of the child-bearers.

These populations inhabit areas of very low density, so low that it is almost inconceivable to the modern urban dweller. The idea of walking for two weeks and never seeing another individual is true solitude. If you did see someone, it would be a member of your immediate household or local band. However, even these small populations are not completely isolated. Occasionally, one of these populations meets another (Fig. 6.1). When this happens a complex set of interactions takes place. There may be immediate withdrawal, competition for resources, warfare, or trade and exchange. This study will be concerned with a model which addresses all but the first alternative. Each alternative is a type of interaction and thus I call my model a model of interactive growth.

My simulation model has several features. First, there are four major groups of parameters. Each is an input entered prior to running the simulation. They are the initial sizes of the populations, the initial growth rates, the competition

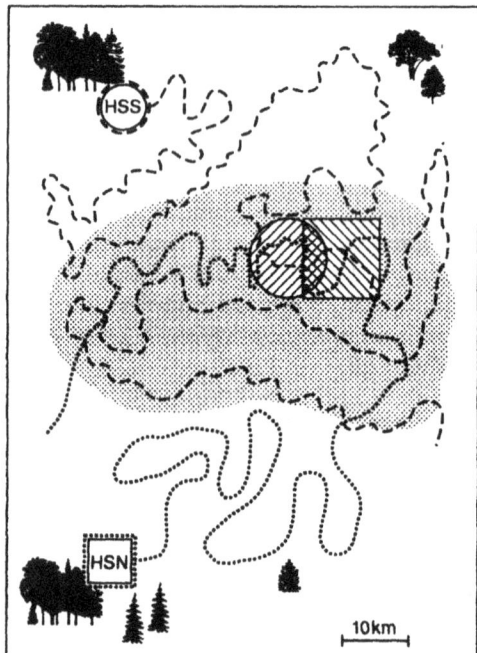

Figure 6.1 The model's scenario: populations of modern humans and Neanderthals pursuing foraging strategies within overlapping territories. *Key:* HSS = *Homo sapiens sapiens*; HSN = *Homo sapiens neanderthalensis*.

10km

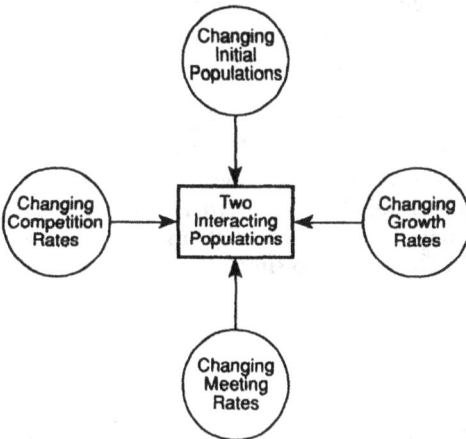

Figure 6.2 The initial parameters for the interactive growth model.

or replacement rates, and the probability that the two populations come into contact (Fig. 6.2). The model positions both populations according to their respective initial sizes and growth rates. These values will change interactively as each population grows and declines. Second, the growth functions may be applied to as many populations as the simulator is interested in studying. In my case, I will limit this study to two populations. Each will be considered as an example of how the growth of many small populations might take place. This limit of two populations creates a highly simplified world which brings out the similarities and the differences in the populations. Third, the model allows the dependency of the two populations to vary. Within the confines of the model, it is possible for the two populations to be totally independent of each other. On the other hand, it is also possible that one population is dependent upon the resources of the other, to whatever degree, or that both are dependent upon the resources of each other. Finally, the model allows one to census the two populations at any time.

The growth functions of the populations are standard growth equations which operate on the entire population. The model is modular and it is possible to use age and sex-specific growth rates as well as stable population equations. However, these are separate topics and are discussed in detail elsewhere (Zubrow 1989). In addition to the size and growth rates of the two populations, the initial inputs included the probability that members of one population will meet members of the other population. It is assumed that the interaction between the two populations is direct. By this I mean that the members of the two populations meet or they are in direct competition. If not, there is no interaction and the populations grow independently. For competition to occur the populations do not actually have to meet. They may compete serially for the same resources or require the same land or water. For example, one population may enter an area, harvest the game, then leave. When the second population arrives in the same area, their harvest has been diminished. What is not allowed for is indirect competition. One population may reduce the resources in their immediate area; this in turn may lower the resources of an adjacent area. This reduction affects the size of the second population. In short, competition which operates through adjacency is not modelled.

The initial rate of replacement is another parameter. What happens when the members of the two populations meet is modelled. Possibilities include the complete or partial replacement of the members of one population in the resource area by members of the second population. Alternatively, the populations may meet, compete, and remain in a position of status quo. The size and rate of the replacement function allows one to simulate the full range of replacement.

Results from the model

As of the time of writing I have simulated more than 300 variations of these parameters. A good indicator of the demographic viability of a population is the number of generations to extinction. If a population does not become extinct in the first 100 generations I consider it successful. In these first runs I set the maximum limit for the number of generations to be simulated at no more than 200. I also set the size of the initial populations as very small, usually between two and 200. In almost 60 per cent of the cases simulated, one of the two populations survived for more than 100 generations. In only three simulations were the Neanderthal populations able to survive for over 100 generations. In comparison, in over 20 simulations the *H. sapiens sapiens* populations were able to survive over 100 generations.

I have rerun the first 50 simulations and added 250 more variations, raising the total number of variations to over 300. Additionally, I have increased the number of generations to 500. I have also expanded the range of initial populations, growth rates, replacement rates, and meeting rates. The results both confirm and elaborate the original conclusions and so I will emphasize these new results. The simulations were run varying one parameter and holding the others constant. Each initial population was allowed to range from one to 6400 individuals. The population growth rates were allowed to vary far beyond reality. They could and did take on any value from 0.000 to 0.050. To interpret these values so that they are not just sterile figures, one should remember that 0.020 would be a 2 per cent annual increase. If this rate was applied constantly, the population would double in 35 years, increase four times in 70, and be slightly greater than eight times the original population in slightly more than a century. The meeting rate varied also from 0.001 to 0.500. This means that of all possible occasions when interactions could occur, the populations actually met and interacted from one out of 1000 times to every other time. The replacement rates were also run between 0.001 and 0.5. At 0.001 in every 100 interactions a replacement took place. Similarly, 0.500 means in one out of every two interactions a replacement took place. A member of one population replaced a member of the other population. This replacement occurs in the context of competition within the localized resource system.

The window for successful Neanderthal survival is very small. As we will see in the following discussion, Neanderthal extinction almost always occurs between 100 and 250 generations: that is, between 2500 years and 7500 years. It can be as short as 30 generations and as long as 350 generations.

However, before disclosing all of the conclusions, I wish to discuss the results in a systematic manner. What I propose to do is to examine the results of varying one parameter at a time. Then I will discuss the simulations which use parameters based upon ethnographic analogy and epipalaeolithic values.

In the following figures the parameters for the simulations were set to standard settings. Then each parameter was varied while the others were held constant. These settings were initial population of *H. sapiens sapiens* equals ten; initial population of *H. sapiens neanderthalensis* equals 100; initial growth rate of *H. sapiens* equals 0.010; initial growth rate of Neanderthals equals 0.010; initial meeting rate equals 0.010; and initial replacement rate equals 0.010.

Figure 6.3 shows the growth of the interacting Neanderthal and modern *H. sapiens* populations when one increases the initial size of the contacting

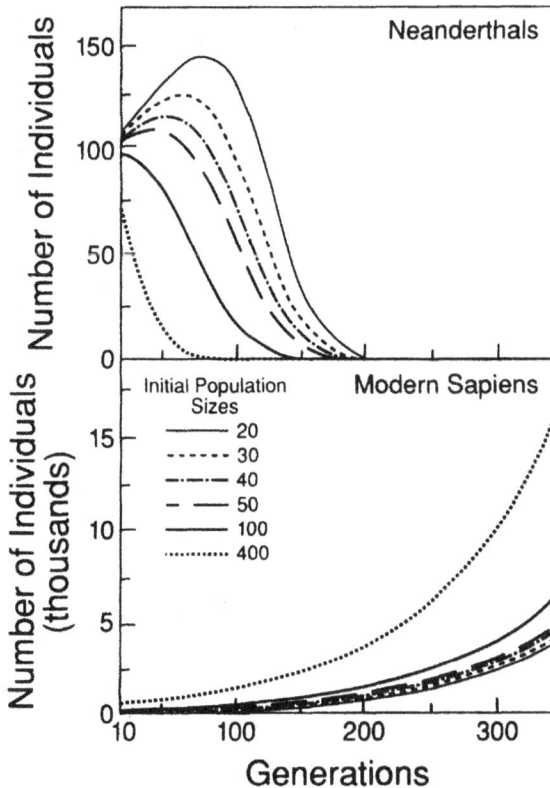

Figure 6.3 Varying the initial population sizes of modern *sapiens*. The parameters are set at initial Neanderthal population equals 100; initial modern *sapiens* equals 20, 30, 40, 50, 100, 400; initial Neanderthal growth rate equals 0.01; initial modern *sapiens* growth rate equals 0.01; the meeting rate equals 0.01; and the replacement rate equals 0.01.

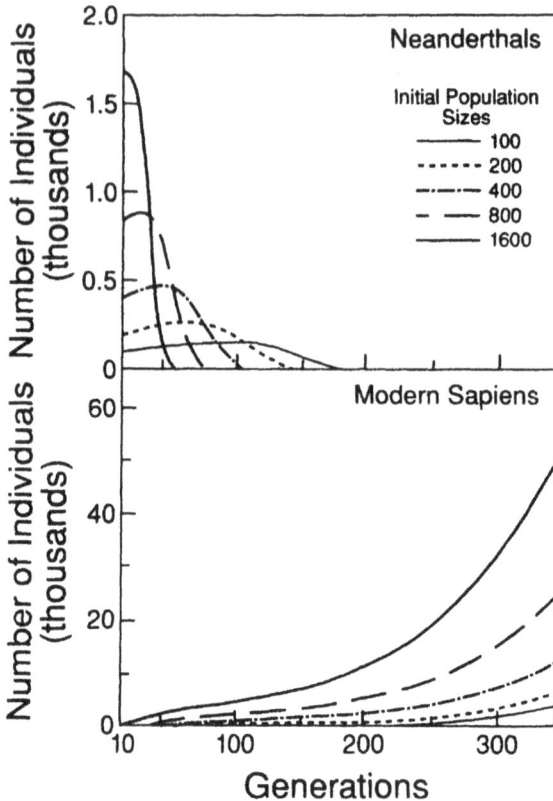

Figure 6.4 Varying the initial population sizes of Neanderthals. The parameters are set at initial Neanderthal population equals 100, 200, 400, 800, and 1600; initial modern *sapiens* equals 10; initial Neanderthal growth rate equals 0.01; initial modern *sapiens* growth rate equals 0.01; the meeting rate equals 0.01; and the replacement rate equals 0.01.

population of modern *H. sapiens*. The upper graph shows the growth of Neanderthals; the lower graph shows the growth of modern *H. sapiens*. Each family of curves represents the change resulting from varying the initial *H. sapiens* population from 20 to 30, 40, 50, 100, and 400. There are several obvious descriptive generalities which should be noted. In all cases the Neanderthals become extinct in less than 200 generations while the *H. sapiens sapiens* grow in a more or less logarithmic function. The rapidity of *H. sapiens* growth is directly related to the size of the initial *H. sapiens* contacting population. The swiftness of Neanderthal extinction is inversely related to the size of the initial *H. sapiens* population. There is a threshold between 100 and 150 for the initial *H. sapiens* contacting population. If this population is above the threshold, the Neanderthal population simply decreases and becomes extinct. This occurs between 70 and 150 generations. If, on the other hand, the number of contacting *H. sapiens* is less than this threshold, both populations grow for a period after contact and it is only later that the Neanderthal populations begin to decline as the growth of the modern *H. sapiens* overtakes them. I call this the 'contact threshold'.

Conversely, Figure 6.4 shows the growth of the interacting populations when I vary the size of the initial Neanderthal population from 100 to 1600. The larger the initial Neanderthal population, the shorter the time to extinction. Thus the Neanderthal population of 1600 becomes extinct in 50 generations while the Neanderthal population of 100 becomes extinct in about 200 generations. The reason this occurs is that the larger Neanderthal population creates the potential for a much larger number of contacts. Thus one could suggest that at time of contact it actually would have been maladaptive if the Neanderthals were in larger groups. The modern *H. sapiens* population grows again more or less logarithmically and with a rate which is directly related to the size of the initial Neanderthal population. Changing the size of the initial *H. sapiens sapiens* population actually causes somewhat less growth marginally than does changing the size of the Neanderthal population. For example, a change from 100 to 400 initial *H. sapiens* results in a change from approximately 6500 to 16 000 *H. sapiens* at the 350th generation. A 400 per cent increase in the initial population results in a 250 per cent increase by the final generation. On the other hand, changing the initial Neanderthal population from 100 to 400 results in an increase in *H. sapiens* at the 350th generation from 3500 to 13 000 or a resultant 370 per cent increase.

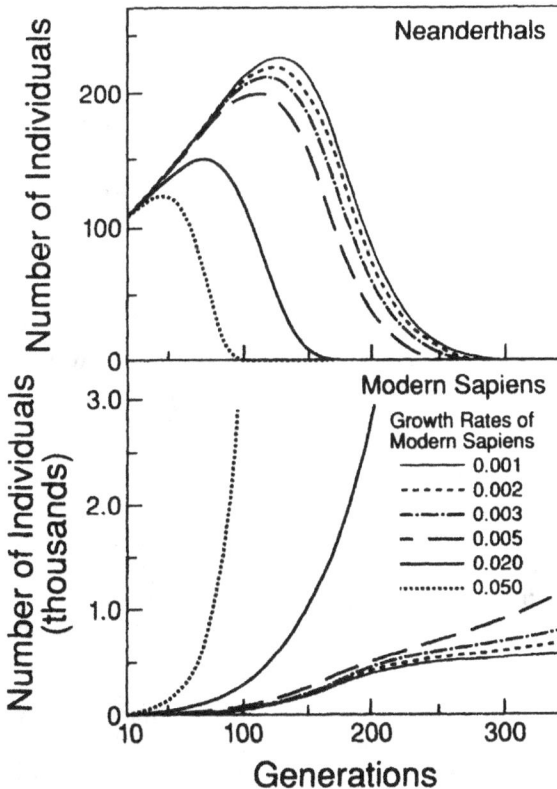

Figure 6.5 Varying the initial growth rate of modern *sapiens*. The parameters are set at initial Neanderthal population equals 100; initial modern *sapiens* equals 10; initial Neanderthal growth rate equals 0.01; initial modern *sapiens* growth rate equals 0.001, 0.002, 0.003, 0.005, 0.020, 0.050; the meeting rate equals 0.01; and the replacement rate equals 0.01.

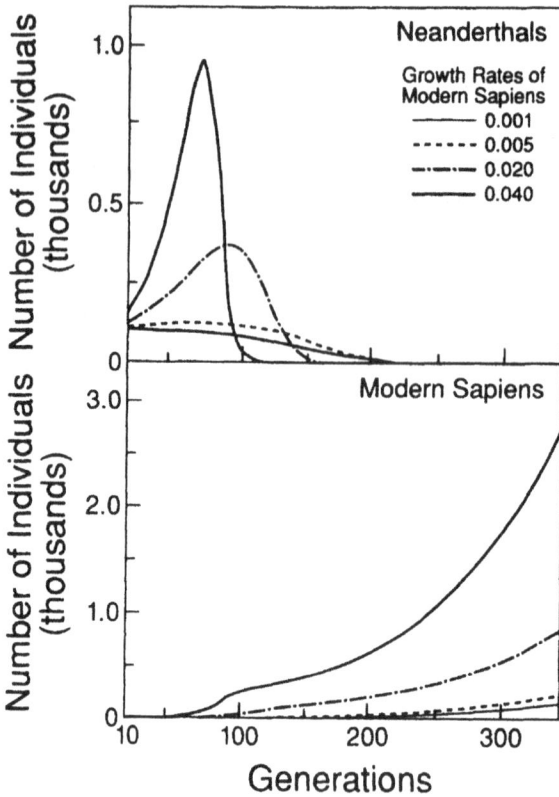

Figure 6.6 Varying the initial growth rate of Neanderthals. The parameters are set at initial Neanderthal population equals 100; initial modern *sapiens* equals 10; initial Neanderthal growth rate equals 0.001, 0.005, 0.020, 0.040; initial modern *sapiens* growth rate equals 0.01; the meeting rate equals 0.01; and the replacement rate equals 0.01.

Figure 6.5 depicts the interaction when I vary the growth rate of the contacting modern *H. sapiens*. The growth rate ranges from 0.001 to 0.050 in this graph. As one expects, increasing the growth rate of the modern *H. sapiens* is inversely related to the rapidity of Neanderthal extinction as well as directly related to modern *H. sapiens* growth. Neanderthal extinction may be as rapid as 100 generations and as slow as 300. After the initial contact both populations grow. For the Neanderthals they continue to grow for approximately half of their postcontact existence. Thus, when the contacting population grows rapidly at 0.050, the growth period of the Neanderthals is 50 generations. The decline from approximately 120 individuals to extinction takes approximately the same number of generations. At lower growth rates the Neanderthal population grows for longer periods and declines for a larger number of generations. There is an important threshold in modern *H. sapiens* growth. It occurs at about 0.010. If one examines the lower graph the curves from 0.001 to 0.005 show a logistic form of growth. By this I mean the growth is relatively slow for the first 100 generations, becomes more rapid for the second 100 generations, and slows down again for the third 100 generations. This corresponds to a model which would suggest a period of successful adaptation, then rapid adaptive radiation, and finally another successful adaptation. At the higher growth rates of 0.020 or 0.050 the modern *H. sapiens* just take off in an almost logarithmic growth pattern. I call this threshold the *sapiens* growth threshold.

Figure 6.6 is the corresponding variation of the Neanderthal growth rates. The growth rates vary from 0.001 to 0.040 in this illustration. The Neanderthals follow the same patterns as we have noted before. Extinction takes place between between 100 and 250 generations or less than 10 000 years. Neanderthal growth rates are inversely correlated to Neanderthal survival. The maladaption of rapid growth is clear. A growth rate of 0.040 results in a population of almost 1000 in 70 generations. Extinction, however, occurs 40 generations later. The results are only slightly less dramatic with rates of 0.020. Once more there is a threshold. Its character is only sketched in this graph. However, if the growth of the Neanderthals is less than 0.005, they do not grow after contact. Contact by the modern *H. sapiens* populations simply rings the death knell of the Neanderthals. It is, however, a long concert taking more than 150 generations. The growth rate of the Neanderthals is directly related to the growth of the *H. sapiens*. The greater the Neanderthal growth, the greater the resultant *H. sapiens* population. The growth is relatively slow. It is not until more than 50 generations have occurred that one can begin to pick out significant differences in the numbers of modern *H. sapiens*. This is partly a result of the scale, but not entirely. Previously, at similar scales, differences in initial populations and growth rates could be determined. One should also note, in passing, that if the Neanderthal growth rate is high enough, one has a logistic curve in the growth of modern *H. sapiens*. After the second plateau, or more accurately the quasi-plateau caused by a decreasing growth rate, the growth picks up significantly and then continues to grow logarithmically.

Briefly, an increase in Neanderthal growth rates from 0.001 through 0.005 to 0.020 results in an eightfold increase in modern *H. sapiens* while the same increase in *H. sapiens* results in first a doubling and then an additional fivefold increase.

Figures 6.7 and 6.8 are very similar. Each represents the changes caused by decreasing their respective parameters: that is, the meeting rate and replacement rate respectively. The demise of the Neanderthals takes place in approximately 250 generations in each of these cases. As the meeting rate and the replacement rate increase, the time to extinction becomes shorter. For the meeting rate there is a threshold between 0.003 and 0.005. If the value is less than 0.003, the Neanderthal population grows before becoming extinct. If more, then the population rapidly becomes extinct without any growth. The threshold for the replacement rate is approximately 0.010. If the competition rate is greater than this value, extinction takes place without any preliminary increase in the Neanderthal population. Although structurally similar in that the shapes of the resultant graphs are the same, there is a quantitative difference between the meeting and the replacement rates. It takes a smaller change in the meeting rate than the replacement rate to create the same decrease in the time for extinction.

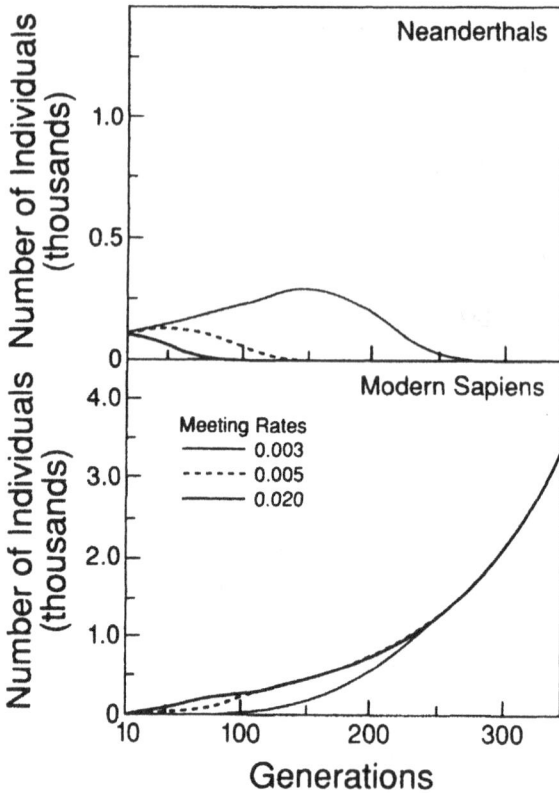

Figure 6.7 Varying the meeting rate for the populations. The parameters are set at initial Neanderthal population equals 100; initial modern *sapiens* equals 10; initial Neanderthal growth rate equals 0.01; initial modern *sapiens* growth rate equals 0.01; the meeting rate equals 0.003, 0.005, 0.020; and the replacement rate equals 0.01.

By now the reader must be crying 'enough'. We have a good idea of how the parameters cause changes in the prehistoric populations. However, there is a limit to speculation without relating it to what is actually known about specific ethnographic and prehistoric populations. In order to replace the reader's feet firmly on the terra firma of anthropological reality, I ran a series of simulations using ethnographic and prehistoric rates. There are of course a considerable number of ethnographic examples which could be used. There are also a great number of assumptions and stretches of imagination that are necessary to use such data. For this chapter I will report on only one set of three cases. I culled the following data from !Kung bushmen ethnographies and demographic studies (Howell 1979, Lee 1972a, 1972b). I set the number of members in the contacting population of *H. sapiens sapiens* as 20, 50, and 500. This corresponds to the range of the ethnographic extended household of 20–50 and to the full band size of approximately 500. I set the Neanderthal population to 500, or the band size. The growth rates for both populations were set to the ethnographic values of 0.0026. The meeting and replacement rates were set at 0.010.

Figure 6.9 illustrates the resulting population curves for these ethnograph-

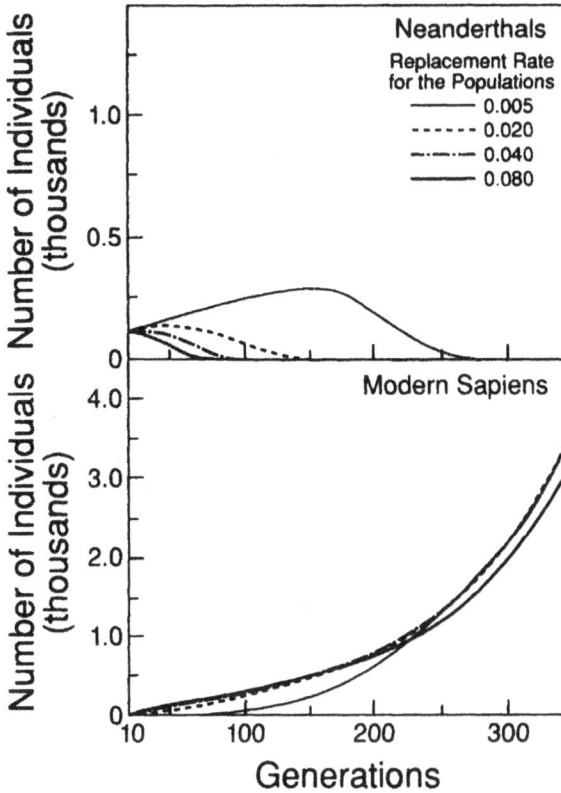

Figure 6.8 Varying the replacement rate for the populations. The parameters are set at initial Neanderthal population equals 100; initial modern *sapiens* equals 10; initial Neanderthal growth rate equals 0.01; initial modern *sapiens* growth rate equals 0.001, 0.002, 0.003, 0.005, 0.020, 0.050; the initial meeting rate equals 0.01; and the initial replacement rate equals 0.005, 0.020, 0.040, 0.080.

ically grounded populations. In all three cases extinction occurs prior to 150 generations. Indeed, when one population meets another coming down a foraging path – that is, band meets band rather than household meets band – the extinction takes place remarkably quickly, in only 50 generations. The growth curves for the replacing population, the modern *H. sapiens sapiens*, have no surprises for us. In all three cases there is rapid growth for the first 30 to 50 generations; then the growth continues, but it is a slower, indeed almost constant, rate.

If one tries to estimate real prehistoric population growth rates, one is entering a very difficult and speculative area. There are not a lot of data to rely upon and what there are have been beset by problems. These problems are not created by the analyst. Rather they are the result of limited samples, poor preservation, and the difficulty of the task. Acsadi & Nemeskeri (1970) have presented one series of data. They are not the only ones, nor are they necessarily the best, but they are well known and appear to be reasonable. If one takes the epipalaeolithic, the best of their earlier sequences, the parameters are set so that the initial populations are both 185 (which was the size of their skeletal populations). The growth rates for both populations are 0.00013. The meeting and replacement rates are 0.01. These latter rates were chosen because they were reasonable and convenient. They were not based upon specific anthropological data. Figure 6.10 shows these results. They conform with the general pattern we have seen in the previous ethnographic cases. Extinction takes place for the Neanderthal population in the first 150 generations. Growth for the modern *H. sapiens* is rapid and then slows down to a linear form.

If you review all of the above, there are several generalities worth emphasizing. First, no matter which parameter one varies or relaxes, the Neanderthal population goes extinct. Usually, the time to extinction is in the neighbourhood of 150 generations. Second, the modern *H. sapiens* populations are particularly hardy and the question is far more frequently how rapid is their growth rather than how long to time of extinction. Third, there are two demographic regimes which can be separated. In one of these regimes the Neanderthal populations continue to grow after initial contact for several generations. Then it appears that the processes of competition, replacement, and the increasing numbers of modern forms overwhelm the Neanderthal growth. In the other regime there is simply decline and extinction after contact. It may be slow and then occurring at an increasing rate, or it may be

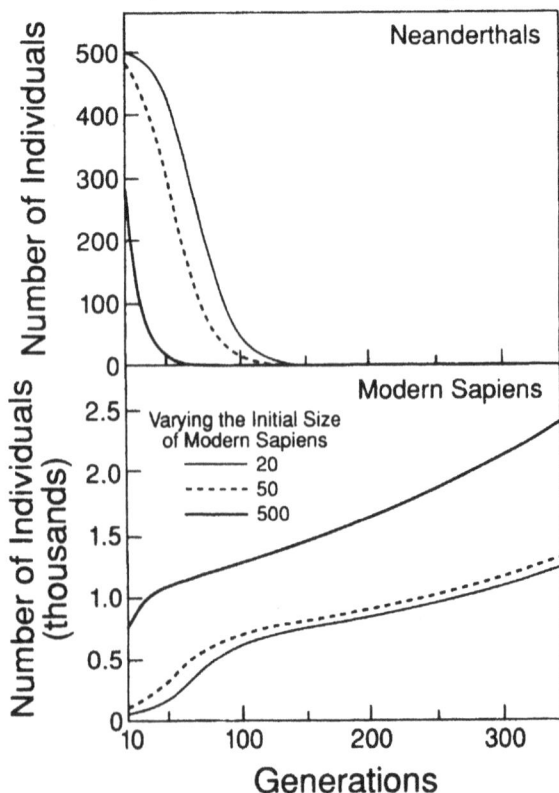

Figure 6.9 Simulations based on ethnographic analogy. The parameters are set at initial Neanderthal population equals 500; initial modern *sapiens* population equals 20, 50, 500; initial Neanderthals and modern *sapiens* growth rates equal 0.0026; initial meeting and replacement rate equals 0.010.

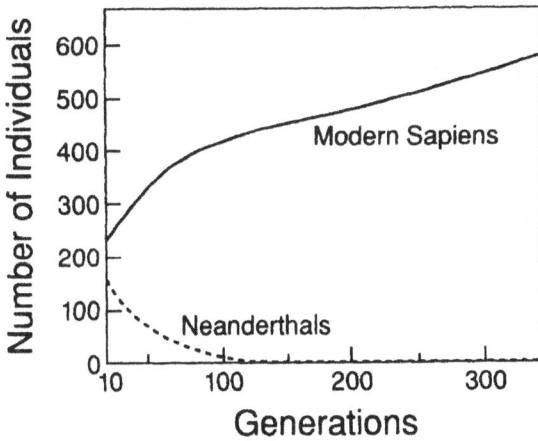

Figure 6.10 Simulations based on epipalaeolithic rates. The parameters are set at initial Neanderthal and modern *sapiens* populations equal 185; initial Neanderthal and modern *sapiens* growth rates equal 0.00013; and the initial meeting and replacement rate equals 0.01.

fast. But the Neanderthals never seem to be able to maintain any growth after contact. These two regimes are separated by a series of thresholds or threshold values for the parameters. One could say that all other things being equal, if the contacting modern *H. sapiens* are a population greater than 150, or have a growth rate greater than 0.010, or if the competition rate is above 0.003, or if the meeting rate is 0.010, then the Neanderthal population enters this second regime and simply declines.

Conclusions

In this chapter I have briefly surveyed some of my ongoing research on simulating interacting prehistoric populations. I have developed a simulation model based on complex interactive growth. It shows that under many different demographic and interactive variations, Neanderthal survival was impossible. The demographic window which could have made it possible was quite improbable – it requires unreasonably low sizes and growth rates for the populations of *H. sapiens sapiens*. Even this was insufficient for Neanderthal survival. Survival would have additionally required very low replacement and interaction rates. In general, Neanderthal continuation was more prolonged in competitive situations where both populations were small. It would appear that one advantage the modern forms had was ability to reach more rapidly a form of 'stable' growth. Needless to say, far more may be accomplished with these models as they become more sophisticated and more simulation runs are completed.

References

Acsadi, Gy & J. Nemeskeri 1970. *History of human life span and mortality*. Budapest: Akamemiai Kiado.
Howell, N. 1979. *The demography of the Dobe !Kung*. New York: Academic Press.

Lee, R. B. 1972a. Population growth and the beginnings of sedentary life among the !Kung bushmen. In *Population growth: anthropological implications*, B. Spooner (ed.), 329–42. Cambridge, Ma.: MIT Press.

Lee, R. B. 1972b. !Kung spatial organization. An ecological and historical perspective. *Human Ecology* 1, 125–47.

Zubrow, E. B. W. 1989. The demographic modelling of Neanderthal extinction. In *The human revolution: behavioural and biological perspectives in the origins of modern humans*, P. A. Mellars & C. Stringer (eds), 212–31. Edinburgh: Edinburgh University Press.

7 *Aboriginal fossil hominids: evolution and migrations*

PHILLIP J. HABGOOD

The earliest evidence for human occupation of Sahul, the combined landmass of Australia, New Guinea, and Tasmania, is at least 40 000 BP (Groube *et al.* 1986, Jones 1989, Nanson *et al.* 1987, Pearce & Barbetti 1981, White & Habgood 1985, White & O'Connell 1982). The sites in question (Huon, Upper Swan, Lake Mungo, Keilor, and Cranebrook Terrace) are located in the northeastern, southwestern, and southeastern parts of the continent

Figure 7.1 Map of Sahul showing maximum low sea level and major late Pleistocene sites.

(Fig. 7.1). So as to allow for the movement from the most probable entry point into the continent – the northwest (Birdsell 1977) – to these dispersed sites, the most widely quoted date for the initial migration(s) to Sahul is around 52 000 BP, when there was a major glacio–eustatic lowering of sea levels (Chappell 1976, 1982). However, the initial entry into Sahul could have been significantly earlier (Jones 1989).

A major debate still rages as to who these colonists were. Three major explanations have been postulated. Two are based on the premise of a number of migrations by morphologically different groups which subsequently inter-bred, while the other contends that the Australian Aborigines migrated from a single biological homeland (see Kirk & Thorne 1976 for references).

Birdsell's (1979) trihybrid theory, based on studies of contemporary Abori-gines, postulates three waves of colonists into Australia. The first wave comprised the Oceanic Negritos, whose remnants he saw in Tasmania and the rainforests of northeastern Australia. No geographic homeland for the Oceanic Negritos has been specified by Birdsell, but he does see them as being present in other areas, such as the highlands of New Guinea, the Andaman Islands, parts of the Malay Peninsula, and on some of the Philippine islands (Birdsell 1977). The second wave comprised the Murrayians, who are linked with the Ainu and who displaced the Oceanic Negritos from most of the continent. Birdsell's final and most recent wave of colonists comprised the Carpentarians, who possibly came from India. The morphologically variable Australian Aborigines were seen by Birdsell as a hybrid of all three groups.

This trihybrid explanation has some major problems. Studies of prehistoric Tasmanian crania (Pardoe 1984, Thorne 1971b) suggest that they are ' . . . variants of a southern Australian population, based on a morphology existing about the time of Tasmania's connection and subsequent separation from the mainland' (Thorne 1971b, p. 319). A study of Queensland crania failed to distinguish between those from the northwestern rainforests and those from the rest of Queensland (Larnach & Macintosh 1970). Also, the Carpentarians seem to be the result of recent contact between Aborigines along the northern Australian coast and Macassan trepang fishermen from Indonesia and Papuan traders (Larnach & Macintosh 1970, Thorne 1971b). An Ainu link for the Murrayians has also been challenged (Yamaguchi 1967).

A dual-source explanation based on the differentiation of the late Pleis-tocene and earlier Holocene Australian skeletal material into two distinct morphological types, one 'robust' and the other 'gracile', has been proposed by Thorne (1971a, 1976, 1977, Thorne & Wilson 1977, see also Freedman & Lofgren 1979). The 'robust' type, as typified by crania from Kow Swamp, Cohuna, Coobool Creek, Talgai, Mossgiel, and Cossack, is relatively low and rugged, with flat, receding frontal bones, marked postorbital constriction, large supraorbital tori and occipital tori, moderate gabling of the thick cranial vault, broad prognathic faces, and large palates, mandibles, and teeth. The 'gracile' type, as exemplified by material from Lake Mungo, Lake Tanou, Lake Nitchie, and Keilor, has high, rounded, and in general more modern-looking crania, with thin vault bones, expanded frontal and temporal squama,

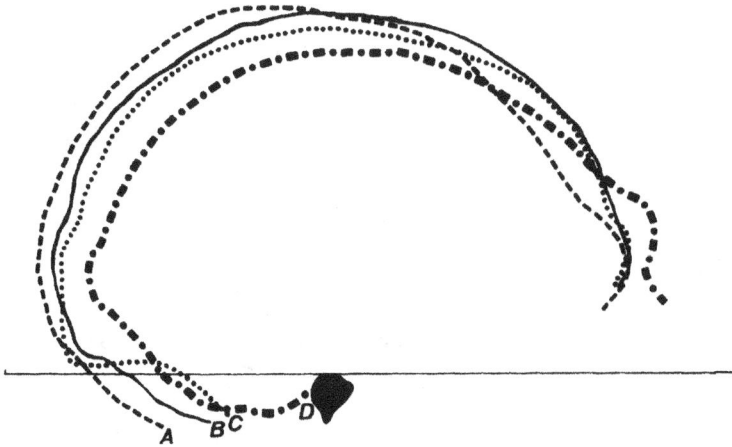

Figure 7.2 Midsagittal cranial contours of Australian Aboriginal crania orientated on the Frankfurt plane (after Freedman & Lofgren 1979). *Key*: A = Cohana; B = Keilor; C = Lake Nitchie; D = Kow Swamp 1.

slight brow-ridge development, lightly constructed nonprognathic facial regions, and relatively small palates, mandibles and teeth. At present, the earliest date for the 'robust' type is approximately 14 000 BP at Kow Swamp, whereas the 'gracile' Lake Mungo 3 skeleton has a date of at least 30 000 BP. Thorne (1977) explained these two types as being the result of two morphologically distinct and chronologically separated groups entering Australia. The 'robust' type, which display 'the mark of ancient Java', came from Indonesia, while the 'gracile' type, which have 'the stamp of ancient China', came from East Asia (Thorne 1977, see also Freedman & Lofgren 1979). Interbreeding between these two groups is thought to have eventually led to the modern Australian Aboriginal morphology.

As with Birdsell's trihybrid explanation, Thorne's dual-source hypothesis has some major problems. Recent multivariate analyses of cranial measurements have shown that the late Pleistocene and early Holocene Australian crania are more similar to each other than they are to either East or Southeast Asian crania (Habgood 1985, 1986b). Also, placing of individual crania into one or other of the two morphological extremes is more difficult than Thorne's hypothesis would suggest. Thorne places the Lake Nitchie cranium in his 'robust' group (Thorne 1977), based solely, one assumes, on its large size, yet examination of the cranial measurements and midsagittal contour suggests a strong similarity with the Keilor cranium, one of Thorne's 'gracile' types (Fig. 7.2, Freedman & Lofgren 1979). Most workers would place Lake Nitchie in Thorne's 'gracile' group (Macintosh 1971, Howells 1973, Freedman 1985, Freedman & Lofgren 1979, Habgood 1985, 1986b). What the individual crania display is a range of morphological forms (Fig. 7.2), not two extremes as Thorne suggests. This range is well illustrated at the two sites with large samples, Kow Swamp and Coobool Creek, which 'include individuals at the opposite ends of the morphological spectrum, and indeed most of the specimens at these sites represent the intermediates between these extremes' (Wolpoff et al. 1984, p. 445). For example, Wolpoff (1980) identified a similarity in the development of the frontal and supraorbital regions between Lake Mungo 3 and some of the more gracile Kow Swamp males such as KS 14 and KS 15, and thought the Lake Mungo 1 cranium resembled the more complete female crania from Kow Swamp, such as KS 4 and KS 16. Webb (1989), however, does not feel that the degree of gracility of Willandra Lakes hominids such as Lake Mungo 1 (his WLH 1) is matched by females from Kow Swamp.

Finally, some of the variation within the late Pleistocene and early Holocene cranial material appears to be due to cranial deformation. Brown (1981) has demonstrated that features, including a flat frontal bone and a prebregmatic eminence, which are said to be typical of the 'robust' type (Thorne 1976), are likely to be the result of deformation caused by cranial pressing. This form of cranial deformation would allow a great deal of variation in the amount of deformation and associated effects on the crania (Brown 1981). What has been produced is a gradation from skulls, such as Kow Swamp 5 and 7, Cohuna, and Coobool Creek 1, 49 and 65, which display marked deformation, to those, such as Kow Swamp 1, 8 and 9, and Mossgiel, which display little or no deformation.

After a recent study of the hominid sample from the Willandra Lakes (the series numbers from WLH 1–135), Webb (1989) concluded that the morphological range was too great to be encompassed within a single morphological population. Like Thorne, he argued that there was a 'robust' group and a 'gracile' group that were the result of separate migrations. Instead of arguing for separate geographical homelands for the two types, he proposed that they both came from the same area, Indonesia, but were separated by a considerable period of time, during which gracilization occurred. That is, the 'robust' type entered Australia first and was later followed by the 'gracile' type that had subsequently developed in Sunda. By the late Pleistocene the Australian population reflected the wide range of morphological variation produced by the intermixing of the two types. He found it difficult to decide whether the 'gracile' type constituted a second population or just a link in the chain of human migrations to Australia and even postulated that the 'gracile' type may be an indication of the emergence and spread throughout parts of Sunda and Sahul of precursor populations that eventually gave rise to the smaller human phenotypes such as the modern 'negrito' stocks.

It is hard to argue against his propositions because of the fragmentary condition of most of the Willandra Lakes hominid material and the lack of chronological control for the sample (it may represent a very long period of time). Also, he does not explain in any detail the mechanisms involved in the gracilization of the Sunda population, especially when one considers the robust nature of late Pleistocene/early Holocene material from other regions, such as North Africa (Anderson 1968, Greene & Armelagos 1972). He argued that the Willandra Lakes hominid sample was made up of two types ('gracile' and 'robust') but does not explain how the two types could live in the same region at the same time without interbreeding and the range of variation decreasing. Finally, he does not explain why his widespread 'negrito/gracile' type was genetically swamped by the 'robust' type when the two groups interbred to produce the late Holocene and modern Australian Aborigines. This point is especially important if the two types coexisted in the Willandra Lakes over a long period of time, as he infers.

Brown (1987, 1989) has also argued that although variation is present at an individual level, there is a consistent Australian Pleistocene morphology, not two separate morphologies.

The third explanation, the homogeneity hypothesis, suggests that the Australian Aborigines are the result of migration from a single biological homeland. Proponents of this explanation have studied both living Aborigines (Abbie 1963, 1968) and skeletal material (Macintosh 1971, Howells 1973, 1976, Macintosh & Larnach 1976, Habgood 1985, 1986b, Brown 1987, 1989), and have concluded that the data suggest a homogeneous founding population for Australia. It should be noted that both Abbie and Howells regarded the Tasmanians as Melanesian and so different to the mainland Aborigines, but as we have seen, prehistoric Tasmanian Aboriginal crania are mainland Australian in their affinities (Pardoe 1984, Thorne 1971b).

What we have in Australia is a corpus of late Pleistocene and early Holocene skeletal material that displays a continuum of cranial forms across a large range of morphological variation (see Habgood 1985, 1986a, 1986b). As Macintosh & Larnach stated, the various fossil crania are 'equal representatives ranged towards (because they fall inside the extremes of the range) either end of a continuum of a single population' (1976, p. 114).

The homogeneity explanation proposes that the morphological variation evident in the late Pleistocene and early Holocene Australian skeletal material was caused by genetic processes, and not due to subsequent migrations by morphologically distinct groups from different geographical homelands and biological sources. This does not necessarily mean that more than one migration from the original source area could not have occurred. The additional colonists could have come from the same biological stock as those that preceded them, or have come in sufficiently small numbers so as not to add substantially to the genetic-morphological make-up of the continental population.

At present this explanation fits the available morphological data better than either of the other two hypotheses. It is also compatible with the archaeological data, in that the corpus of late Pleistocene Australian stone tools is so similar, even when they are made from different materials or utilized in different environments, that they are grouped into the pan-continental Australian 'core-tool and scraper tradition'. Similarly, the first distinct corpus of rock art in Australia, the Panaramitee style, is found throughout the continent, including Tasmania, and forms a relatively homogeneous stylistic entity (Franklin 1990, Maynard 1979, White & Habgood 1985, White & O'Connell

1979, 1982). The late Pleistocene archaeological record, therefore, also suggests that the continent of Australia was colonized by a group (or groups) with a homogeneous cultural background and, by inference, from a single source area, and not from a multitude of cultural sources.

If we assume that Sahul was colonized by people from one source area, as the current evidence suggests, where was this area? The most likely region is Indonesia, which incorporates the southern part of Sunda Land and the Wallacean islands (Fig. 7.1). Whether the colonists had lived in Sahul for a long period or had recently migrated into the area is open to much debate (see Habgood, in press, Smith et al. 1979, Stringer & Andrews 1988 for discussions of this problem). At present, the only human fossils from Indonesia that could possibly be the ancestors of the earliest inhabitants of Sahul are the Ngandong hominids, which are generally classified as Homo erectus (Coon 1962, Santa Luca 1980, Stringer 1984, Groves 1989, Habgood 1989, in press).

The date of the Ngandong hominid sample remains difficult to ascertain (see Bartstra et al. 1988, Habgood, in press). The hominids were recovered from the upper or 20-m terrace of the Solo river near Ngandong, central Java (Santa Luca 1980, Weidenreich 1951). They were not localized in any particular spot or within a single layer but were irregularly distributed throughout the entire site, and so while they can be regarded as a sample, they do not necessarily represent a single biological population. This terrace of the Solo river, which is referred to as the Notopuro formation, contained abundant, predominantly extant, mammalian fauna (the Ngandong fauna), and so is usually considered to be of Upper Pleistocene age (Santa Luca 1980). This has meant that some scholars, such as Coon (1962), have given the Ngandong hominids an Upper Pleistocene date.

Santa Luca (1980), however, has provided taphonomic evidence to suggest that the Ngandong fauna is a mixed assemblage, with the nonhominid fauna representing a death-assemblage buried after minimal exposure, while the hominid remains appear to have been redeposited into younger levels. This suggestion is consistent with the arrangement of the hominid remains within the terrace. The nonhominid fauna is also well preserved, with complete and articulated vertebral columns and crania together with associated mandibles, whereas the hominid material is fragmented and consists predominantly of calvaria. The hominid sample is composed of only the most durable structures, with the calvaria displaying evidence of surface damage and lacking the facial skeleton, which could be the result of rolling and transportation of the crania by water (Boaz & Behrensmeyer 1976). This pattern is consistent with the suggestion that the hominid sample had been redeposited. It would appear that the Ngandong hominid sample has had a complex depositional history which will make the dating of it very difficult. A middle or late Middle Pleistocene age for the Ngandong hominids would seem a reasonable estimate at present (but see Bartstra et al. 1988 for an earlier dating).

If the concept of regional morphological continuity within Australasia is correct, an idea that is much debated (see Groves 1989, Stringer 1984, Stringer & Andrews 1988, Stringer et al. 1984, Brauer 1984, 1989, Habgood 1986b,

1989, in press), and there is an evolutionary line leading from the early Indonesian *H. erectus* type through to the Australian Aborigines, via the Ngandong form (Weidenreich 1943, Coon 1962, Macintosh 1965, Thorne & Wolpoff 1981, Wolpoff *et al.* 1984), then the earliest inhabitants of Sahul should be advanced Ngandong type *H. sapiens* (assuming a date around 52 000 BP for the initial occupation of the continent).

The discovery of Willandra Lakes hominid 50 (WLH 50) may support this assumption (Flood 1983, Thorne 1984, Webb 1989). A calvaria, some portions of the facial skeleton, and fragmentary postcranial material were found on the surface near Lake Garnpung, north of Lake Mungo. This essentially unpublished hominid is opalized, with all the normal phosphate in the bone being replaced by silicates (Flood 1983). The large and robust calvaria is fully sapient in overall configuration and Australian Aborigine in affinity (Webb 1989). WLH 50 has very thick vault bones. Over 70 per cent of this thickness is made up of diploeic bone (Delson 1985, Flood 1983, Thorne 1984, Webb 1989). The very angular calvaria has a flat and receding frontal, marked temporal crests, a protruding occipital with a well-developed transverse torus, marked angulation between the occiput and the nuchal plane, prominent brow-ridges (especially the medial segment), and maximum parietal breadth located towards the parietal mastoid angle. The fragmentary postcranial elements are also quite large and robust (Flood 1983).

The date of WLH 50 has not been conclusively established. The skeletal material was a surface find and so it is difficult to ascertain to which geological layer it should be equated. Quoting unpublished papers written by A. G. Thorne, Flood (1983, p. 67) recorded that, 'Radiocarbon and trace element analysis indicate a minimum age of 25 000 to 30 000 BP, but the remains are probably much older', while Delson (1985, p. 298) stated that, 'Using an experimental electron spin resonance approach [ESR], the oldest specimen, WLH 50, is much older than 30 000, perhaps something like 60 000 years old.' These dates are preliminary estimates and have not been substantiated in print. Caddie *et al.* (1987) calculated an ESR date of 29 000 ± 5000 years for WLH 50, based on the assumption that the natural dose rate at the site of WLH 50 was the same as that for other sites studied in the area. The crust of calcium carbonate around WLH 50 has provided a minimum radiocarbon date of 15 000 BP (A. G. Thorne pers. comm.). Webb (1989) contends that WLH 50, along with the rest of the Willandra Lakes hominid sample, predates the end of lunette formation at the Willandra Lakes, around 15 000 BP. It is, however, unlikely that WLH 50 is older than 45 000 BP, as this is the estimated date for the bottom of the Mungo Unit, and as yet no archaeological remains have been recovered from the underlying Golgol Unit (Bowler 1976, White & O'Connell 1982).

Thorne (1984, p. 227) declared that WLH 50 was 'much more robust and archaic than any Australian hominid found previously'. He and Wolpoff (1985) have proposed resemblances between early Indonesian crania and WLH 50, especially in the form of the frontal squama, and the skull shape viewed in *norma occipitalis*, with Thorne (pers. comm.) suggesting that the transverse arc of WLH 50 was close to the mean value for this measurement in the Ngandong sample. Elsewhere I have argued that the morphological features that have been suggested to demonstrate a morphological link between Australian crania and earlier Indonesian material are generally characteristic of *H. erectus* and archaic *H. sapiens* throughout the Old World (Habgood 1989, in press). However, a combination of a number of fronto-facial features (a long and sagittally flat frontal bone with a posterior position of minimum frontal breadth, malars with everted lower margins and prominent zygomaxillary tuberosities, and possibly very prognathic faces) did seem to indicate some degree of regional morphological continuity in Australasia (Habgood 1989a, in press). WLH 50 has this combination of features (Habgood 1989a) and so may be evidence for regional continuity in Australasia, especially if it is older than the Lake Mungo material, which does not support morphological continuity with earlier Indonesian hominids (Habgood 1989a, in press). WLH 50 may, therefore, resemble the original colonizers of Sahul.

However, one cannot discount the possibility of some pathological changes to the vault of WLH 50 caused, most probably, by some form of haemoglobinopathy (Brown 1987, Webb 1989, G. E. Kennedy pers. comm.). Although the heavy mineralization of the calvaria interfered with radiographic visualization, X-rays of WLH 50 reveal a somewhat lineal (hair-on-end) arrangement of the diploeic trabeculae around the bregma which is consistent with a diagnosis of severe anaemia. Even if pathological modifications did occur, WLH 50 would have been a very robust cranium and would have presented the combination of morphological features listed above.

A continued biological link with, or the continued influx of genes from, Indonesia is suggested by the morphological similarity between the undated, but probably mid-Holocene, Wajak 1 cranium and the 12 900-year-old cranium from Keilor in southeastern Australia (Coon 1962, Habgood 1985, 1986b, Weidenreich 1943, Wolpoff *et al.* 1984). It is also probable that the dingo, a semidomesticated dog which appears in Australia about 4000 BP, was brought into the continent by people.

We can assume that the initial population of Sahul was relatively small. At this time the northwest coast of the continent would have been cooler, having an average mean annual temperature drop of up to 5°C, and drier, with a decrease in rainfall of possibly 30 per cent, as well as a change in the seasonality of the rainfall and a diminished effect from tropical cyclones (Bowler *et al.* 1976, Deacon 1985, Webster & Streten 1978, Chappell & Grindrod 1983). It was most probably covered by open woodland and savanna, associated with generally low nutrient soils on laterites, while the coasts may have offered little permanence of tenure due to an unstable environment caused by frequent fluctuations in sea levels on the gently sloping continental shelf (Chappell & Thom 1977, Chappell & Grindrod 1983, Deacon 1985, Hope & Hope 1976). These environmental conditions could have reduced carrying capacities and kept extinction rates high and population numbers low (Deacon 1985, Hayden 1972). Population growth might have been slow due to a low survivorship while inhabitants adapted to the new environment and its resources, and because of physiological controls such as infanticide, prolonged lactation, abortion, and abstinence (Hayden 1972).

To be successful, a colonizing group needed to have the genetic capacity to continue the population. McArthur (1976, McArthur *et al.* 1976) conducted a number of simulation models of the chances of long-term survivorship of small founding populations. It was found that the larger the initial group the greater its chance of survival, and that the populations founded by younger adults had better chances of survival. Groups that did not practise monogamy would also have had a substantially better chance of survival. Differential extinction of populations that may have survived for some time would have helped keep population numbers low.

A low continental population during the late Pleistocene fits with the dearth of sites that are dated to this period, even allowing for poor archaeological visibility and a high degree of site destruction (Jones 1989, White & Habgood 1985, White & O'Connell 1982). The early sites are also not rich in archaeological remains, which suggests a very dispersed, low-density population (Deacon 1985). Some areas, such as Tasmania and the southeastern and southwestern corners of Australia, appear to have been refugia during this period. Substantial population growth is not archaeologically visible in Australia until the mid-Holocene (Beaton 1983, Lourandos 1983, Veth 1989).

As groups colonized the Sahul continent they moved into a diversity of terrains and climates. By at least 20 000 BP most parts of Sahul show evidence of habitation or resource exploitation: for example, Koonalda Cave, Lindner, and Puritjarra in arid regions, Kutikina Cave, Beginner's Luck Cave, and Kosipe in cold upland regions, Matenkupkum and Wallen Wallen Creek in coastal environments, and Lake Mungo, Keilor and Upper Swan in riverine/lake environments (Jones 1989, Veth 1989). If population numbers were relatively small, as I have suggested, it is probable that these colonizing groups would have become isolated from each other due to the enormous size of the continent (see White & Habgood 1985, White & O'Connell 1982 for an evaluation of the various theories regarding the colonization of Australia). There are few significant topographical barriers in Australia except the expanse of ocean that now separates Tasmania from the mainland, and the three major sand-ridge deserts of central Australia (Veth 1989). The whole of the arid core of Australia may also have acted as a temporary barrier to human occupation (and possibly movement) during the last glacial maximum (Veth 1989).

If small groups were isolated, one could expect to find unique local artefactual specializations, which could be adaptations to local conditions. The tula adze found in the arid regions of northern Australia, ground stone hatchets from sites in Arnhem Land, northern Australia, and small tools recovered from early sites in southwestern Western Australia (White & O'Connell 1982) may be examples of this form of local adaptation. Also, within the Panaramitee style there are differing regional emphases on motif types although the overall range is similar and the technique of rendering the motifs varies from pecking, abrading, and pounding through to painting

(Franklin 1990). These differences may be the result of, and evidence for, isolation during the late Pleistocene.

Geographical isolation would have been accentuated by the last glacial maximum, centred around 18 000 BP, when one can expect populations to have decreased. Deacon & Thackeray (1984) proposed a model for southern Africa which suggested marked depopulation as a consequence of lowered usable productivity of the environment due to climatic changes at this time. They assumed changes in both population distribution and density, with local population extinctions. Gamble (1983) has documented a similar depopulation of large parts of central Europe at various times during the Upper Pleistocene. Based on a biogeographic model that divides Australia up into refuges, corridors, and barriers, Veth (1989) identified a lack of evidence for the occupation of large tracts of Australia (the corridors and barriers) during the glacial maximum. Also, Hiscock (1984) hypothesizes that increases in discard rates at Colless Creek Cave between 13 500 BP and 17 000 BP, a period of increased aridity in northwest Queensland, could have been caused by a reduction in the territory the occupants utilized and a more intensive utilization of the resources of the well-watered Lawn Hill Gorge complex. Smith (1989) has argued for a general model of reduced foraging territory during the glacial maximum for the arid region of Australia.

Geographical isolation is a reversible phenomenon which in itself does not affect the separated groups (gene pools) but allows other processes to accumulate genetic differences. It is probable that small colonizing groups would not be a representative cross-section of the parent population and so, due to founder effect, would be genetically different. This effect can be accentuated if the group is composed of members of the same family lineage (lineal effect). Utilization of vastly different environments would mean different kinds of selection would have been acting on the small populations. This may have caused genetic frequency variation between groups, because human biological variation is determined by the interaction between the environment and genetic systems. Random genetic changes or mutations, which are the source of new genetic variation within a gene pool, stand a much greater chance of becoming fixed in small populations and causing a random genetic differentiation between the isolated colonizing groups. Genetic drift, or chance fluctuations in gene frequencies, may also cause genetic differentiation. For example, in isolated and possibly polygynous groups a few males may father the majority of the children, having a profound effect on the gene fixation of each generation. Lourandos (1983) has suggested that the late Pleistocene may have been typified by restricted marriage systems, which would have accentuated isolation by further restricting gene flow.

Under this model what we would have had in the late Pleistocene in Australia is small isolated groups scattered throughout the continent developing genetic variation. Genetic variation does not necessarily mean a high degree of morphological variation, but we can assume a certain correlation between the two. Each isolated group, therefore, could have developed unique genetic and morphological combinations which became fixed. These groups could have been relatively homogeneous, internally, but would have differed substantially from other such groups. An example of this may be the differences between the approximately 12 000- to 13 000-year-old Keilor and Talgai crania. During the late Pleistocene and earlier Holocene the continental population of Greater Australia would, under the above scenario, have been genetically and morphologically heterogeneous, while being made up of many small, and relatively isolated, morphologically homogeneous groups. However, as Brown (1982, 1989) has demonstrated, although there is variation, there was a consistent Australian Pleistocene morphology.

Genetic differentiation and the development of unique genetic combinations could have been restrained by gene flow, but isolation would have been limiting its potential effect. It is significant that in areas such as southeastern Australia, where isolation would have been greatly reduced and gene flow high, the sites of Kow Swamp and Coobool Creek have skeletal samples that display a large range of morphological variation and not fixed unique morphologies (Brown 1982, 1987, 1989, Thorne 1976, Wolpoff 1980; Lourandos 1983 takes a different position, see discussion in Pardoe 1988). This may also have been the case throughout the Willandra Lakes system (Webb 1989). During the late Holocene and at the time of European contact, southeastern Australia, especially along the Murray river corridor, is suggested to have had a higher population density than many other regions of Australia (Butlin 1983, Webb 1984). This situation may also have typified earlier periods if the appearance of burial grounds can be taken to imply population increase and/or higher population densities (Pardoe 1988). The snow-fed Murray-Murrumbidgee river system may also have experienced less extreme conditions during the last glacial maximum, thus preventing major depopulation of the region during this period.

Genetic variation caused by isolation and small population numbers, along with cranial deformation (Brown 1981), can account for the morphological pattern evident in the late Pleistocene and earlier Holocene Australian cranial material.

After the last glacial maximum population numbers would have slowly increased. In its simplest terms this population increase (repopulation may be more accurate) would mean more people or groups inhabiting the landscape. Isolation of groups would, therefore, have gradually lessened. This increase is especially evident (archaeologically visible) during the Holocene (Beaton 1983, Lourandos 1983, Veth 1989).

As population numbers and densities increased, gene flow would have been higher, introducing new genetic material and/or changing gene frequencies in the previously isolated groups. The trend would have been for gradually increasing differentiation within the individual groups, which previously had unique morphologies, until all groups displayed similar types and ranges of variation. Unique morphologies would have disappeared and the chance of new ones becoming fixed would have been greatly lessened in possibly larger-sized groups that were no longer isolated.

The genetic and morphological variation within the smaller groups would have increased, whereas the continent-wide range of variation may have decreased in that there would have been a reduction in the occurrence of groups displaying unique morphologies. In this way the overall continental range of genetic and morphological variation may have been reduced from that of the earlier period, whereas the range of variation of individual groups could have increased with the introduction of new variation due to increased gene flow. Lourandos (1983) contends that during the Holocene increasingly widening marriage systems developed, further increasing the gene flow. Under this scenario, the continental population, which would have seemed more homogeneous, would have been made up of morphologically more variable groups. Areas that may have remained isolated longer than the more optimal regions could still have produced unique morphologies well into the Holocene. An example of this could be the Cossack cranium from Western Australia.

There may, however, have been a change in the pattern described above prior to European contact. Groups living in resource-rich areas that supported high population densities appear to have had rigid territorial boundaries and short marriage distances, therefore restricting gene flow, whereas in arid regions with low population densities and unpredictable resources, groups maintained more fluid boundaries and extensive social networks, which would have promoted gene flow (Pardoe 1988, Peterson 1976, 1986, White 1979). These patterns most probably came into existence with repopulation after the last glacial maximum. That is, when groups moved back into and permanently occupied arid regions they took with them extensive social networks, and as population densities reached critical levels in regions such as the Murray river corridor (Webb 1984), more rigid territorial boundaries became necessary so as to maintain control of resources. This change may have influenced the morphological range of more recent Aboriginal crania, but would not affect the pattern of the late Pleistocene and earlier Holocene material.

From the preceding discussion one can see that to account for the large range of morphological variation of the late Pleistocene and earlier Holocene Australian Aboriginal crania it is unnecessary to resort to explanations involving independent migrations by different groups who remained biologically separated for over 30 000 years before interbreeding to produce the modern Australian Aborigines. A more pedestrian and parsimonious explanation proposes migrations by small groups from a single source (geographic and biological) which, due to isolation during continental colonization and demographic variation such as marked depopulation during the glacial maximum, were acted upon by genetic processes, including founder effect, selection, mutation, genetic drift, and varying amounts of gene flow, causing the development of a large range of morphological and genetic variation. Cranial deformation was also a contributing factor. During the late Pleistocene there would have been small and relatively homogeneous groups making up a heterogeneous continental population. With population increase during the mid-Holocene isolation would have decreased while gene flow increased, causing a reduction in genetic and morphological variation. There would, therefore, have been possibly larger and more heterogeneous groups making up a relatively homogeneous continental population during the mid-Holocene.

Acknowledgements

Financial support for the research upon which this chapter is based and to attend the 1986 World Archaeological Congress in Southampton, where an earlier draft was presented, was provided by the Carlyle Greenwell Research Fund, the University of Sydney, and the Australian Institute of Aboriginal Studies. I would also like to thank Natalie Franklin for typing the final draft of the chapter.

References

Abbie, A. A. 1963. Physical characteristics of Australian Aborigines. In *Australian aboriginal studies*, H. Shiels (ed.), 89–107. Oxford: Oxford University Press.

Abbie, A. A. 1968. The homogeneity of Australian Aborigines. *Archaeology and Physical Anthropology in Oceania* 3, 221–31.

Anderson, J. E. 1968. Later Paleolithic skeletal remains from Nubia. In *The prehistory of Nubia*, F. Wendorf (ed.), 996–1040. Dallas: Southern Methodist University Press.

Bartstra, G-J., S. Soeghondho & A. van der Wijh 1988. Ngandong man: age and artefacts. *Journal of Human Evolution* 17, 325–7.

Beaton, J. M. 1983. Does intensification account for changes in the Australian Holocene archaeological record? *Archaeology in Oceania* 18, 94–7.

Birdsell, J. H. 1967. Preliminary data on the trihybrid origin of the Australian Aborigines. *Archaeology and Physical Anthropology in Oceania* 2, 100–55.

Birdsell, J. H. 1977. The recalibration of a paradigm for the first peopling of Greater Australia. In *Sunda and Sahul: prehistoric studies in Southeast Asia, Melanesia and Australia*, J. Allen, J. Golson & R. Jones (eds), 111–67. London: Academic Press.

Birdsell, J. H. 1979. A reassessment of the age, sex and population affinities of the Niah cranium. *American Journal of Physical Anthropology* 50, 419.

Boaz, N. T. & A. K. Behrensmeyer 1976. Hominid taphonomy: transport of human skeletal parts in an artifical fluviatile environment. *American Journal of Physical Anthropology* 45, 53–60.

Bowler, J. M. 1976. Recent developments in reconstructing late Quaternary environments in Australia. In *The origin of the Australians*, R. L. Kirk & A. G. Thorne (eds), 55–77. Canberra: Australian Institute of Aboriginal Studies.

Bowler, J. M. & A. G. Thorne 1976. Human remains from Lake Mungo: discovery and excavation of Lake Mungo III. In *The origin of the Australians*, R. L. Kirk & A. G. Thorne (eds), 127–38. Canberra: Australian Institute of Aboriginal Studies.

Bowler, J. M., R. Jones, H. Allen & A. G. Thorne 1970. Pleistocene human remains from Australia: a living site and human cremation from Lake Mungo, western New South Wales. *World Archaeology* 1, 39–60.

Bowler, J. M., G. S. Hope, J. N. Jennings, G. Singh & D. Walker 1976. Late Quaternary climates of Australia and New Guinea. *Quaternary Research* 6, 359–94.

Brauer, G. 1984. The Afro-European sapiens-hypothesis, and hominid evolution in

Asia during the later Middle and Upper Pleistocene. *Courier Forschungsinstitut Senckenberg* **69**, 145–66.

Brauer, G. 1989. The evolution of modern humans: a comparison of the African and non-African evidence. In *The human revolution: behavioural and biological perspectives on the origins of modern humans*, P. A. Mellars & C. B. Stringer (eds), 123–54. Edinburgh: Edinburgh University Press.

Brown, P. 1981. Artificial cranial deformation: a component in the variation in Pleistocene Australian Aboriginal crania. *Archaeology in Oceania* **16**, 156–67.

Brown, P. 1987. Pleistocene homogeneity and Holocene size reduction: the Australian human skeletal evidence. *Archaeology in Oceania* **22**, 41–7.

Brown, P. 1989. *Coobool Creek*. *Terra Australis* 13. Department of Prehistory, Research School of Pacific Studies, Australian National University, Canberra.

Butlin, N. 1983. *Our original aggression*. Canberra: Australian National University Press.

Caddie, D. S., D. S. Hunter, P. J. Pomery & H. J. Hall 1987. The ageing chemist – can electron spin resonance (ESR) help? In *Archaeometry: further Australasian studies*, W. R. Ambrose & J. M. J. Mummery (eds), 167–76. Canberra: Australian National University Press.

Chappell, J. 1976. Aspects of later Quaternary palaeogeography of the Australian – East Indonesian region. In *The origin of the Australians*, R. L. Kirk & A. G. Thorne (eds), 11–22. Canberra: Australian Institute of Aboriginal Studies.

Chappell, J. 1982. Sea levels and sediments: some features of the context of coastal archaeological sites in the tropics. *Archaeology in Oceania* **17**, 69–78.

Chappell, J. & A. Grindrod 1983. *CLIMANZ: Proceedings of the First Climanz, 1981*. Canberra: Department of Biogeography and Geomorphology, Australian National University.

Chappell, J. & B. G. Thom 1977. Sea levels and coasts. In *Sunda and Sahul: prehistoric studies in Southeast Asia, Melanesia and Australia*, J. Allen, J. Golson & R. Jones (eds), 275–91. London: Academic Press.

Coon, C. S. 1962. *The origin of races*. New York: Alfred A. Knopf.

Deacon, H. J. 1985. How did past climates affect prehistoric people in Australia and South Africa? *The Digging Stick* **2** (2), 5–6.

Deacon, H. J. & J. F. Thackeray 1984. Late Pleistocene environmental changes and implications for the archaeological record in southern Africa. In *Late Cainozoic palaeoclimates of the southern hemisphere*, J. C. Vogel (ed.), 375–90. Rotterdam: Balkema.

Delson, E. 1985. Late Pleistocene human fossils and evolutionary relationships. In *Ancestors: the hard evidence*, E. Delson (ed.), 296–300. New York: Liss.

Flood, J. 1983. *Archaeology of the dreamtime*. Sydney: Collins.

Franklin, N. R. 1990. Explorations of variability in Australian prehistoric rock engravings. Unpublished PhD dissertation, Department of Archaeology, La Trobe University.

Freedman, L. 1985. Human skeletal remains from Mossgiel, NSW. *Archaeology in Oceania* **20**, 21–31.

Freedman, L. & M. Lofgren 1979. Human skeletal remains from Cossack, Western Australia. *Journal of Human Evolution* **8**, 283–99.

Freedman, L. & M. Lofgren 1983. Human skeletal remains from Lake Tandou, NSW. *Archaeologyy in Oceania* **18**, 98–105.

Gamble, C. 1983. Culture and society in the Upper Palaeolithic of Europe. In *Hunter-gatherer economy in prehistory*, G. N. Bailey (ed.), 201–11. Cambridge: Cambridge University Press.

Greene, D. L. & G. Armelagos 1972. *The Wadi Halfa Mesolithic population.* Amherst Research Report 11. Amherst: Department of Anthropology, University of Massachusetts.

Groube, L., J. Chappell, J. Muke & D. Price 1986. A 40 000-year-old human occupation site at Huon Peninsula, Papua New Guinea. *Nature* **324**, 453–5.

Groves, C. 1989. A regional approach to the problem of the origin of modern humans in Australasia. In *The human revolution: behavioural and biological perspectives on the origins of modern humans,* P. A. Mellars & C. B. Stringer (eds), 274–85. Edinburgh: Edinburgh University Press.

Habgood, P. J. 1985. The origin of the Australian Aborigines: an alternative approach and view. In *Hominid evolution: past, present and future,* P. V. Tobias (ed.), 367–80. New York: Liss.

Habgood, P. J. 1986a. A late Pleistocene prehistory of Australia: the skeletal material. *Physical Anthropology News* **5**, 1–5.

Habgood, P. J. 1986b. The origin of the Australians: a multivariate approach. *Archaeology in Oceania* **21**, 130–7.

Habgood, P. J. 1989. The evolution of modern humans: evidence from Australasia seen in a global context. In *The human revolution: behavioural and biological perspectives on the origins of modern humans,* P. A. Mellars & C. B. Stringer (eds), 245–73. Edinburgh: Edinburgh University Press.

Habgood, P. J. In press. *A morphometric investigation into the origin of anatomically modern humans.* British Archaeological Reports.

Hayden, B. 1972. Population control among hunter/gatherers. *World Archaeology* **4**, 205–21.

Hiscock, P. 1984. Preliminary report on the stone artefacts from Colless Creek Cave, Northwest Queensland. *Queensland Archaeological Research* **1**, 120–51.

Hope, J. H. & G. S. Hope 1976. Palaeoenvironments for man in New Guinea. In *The origin of the Australians,* R. L. Kirk & A. G. Thorne (eds), 29–55. Canberra: Australian Institute of Aboriginal Studies.

Howells, W. W. 1973. *The Pacific islanders.* London: Weidenfeld & Nicolson.

Howells, W. W. 1976. Metrical analysis in the problem of Australian origins. In *The origin of the Australians,* R. L. Kirk & A. G. Thorne (eds), 141–60. Canberrra: Australian Institute of Aboriginal Studies.

Jones, R. 1989. East of Wallace's line: issues and problems in the colonization of the Australian continent. In *The human revolution: behavioural and biological perspectives on the origins of modern humans,* P. A. Mellars & C. B. Stringer (eds), 743–82. Edinburgh: Edinburgh University Press.

Kirk, R. L. & A. G. Thorne 1976. In *The origin of the Australians,* R. L. Kirk & A. G. Thorne (eds), 1–8. Canberra: Australian Institute of Aboriginal Studies.

Larnach, S. L. & N. W. G. Macintosh 1970. *The craniology of the Aborigines of Queensland.* Oceania Monograph 15.

Lourandos, H. 1983. Intensification: A late Pleistocene–Holocene archaeological sequence from southwestern Victoria. *Archaeology in Oceania* **18**, 81–94.

McArthur, N. 1976. Computer simulations of small populations. *Australian Archaeology* **4**, 53–7.

McArthur, N., I. W. Saunders & R. L. Tweedie 1976. Small population isolates: a micro-simulation study. *Journal of the Polynesian Society* **85**, 307–26.

Macintosh, N. W. G. 1965. The physical aspects of man in Australia. In *Aboriginal man in Australia,* R. M. Berndt & C. H. Berndt (eds), 29–70. Sydney: Angus and Robertson.

Macintosh, N. W. G. 1971. Analysis of an Aboriginal skeleton and a pierced tooth necklace from Lake Nitchie, Australia. *Anthropologie* 9, 49–62.

Macintosh, N. W. G. & S. L. Larnach 1976. Aboriginal affinities looked at in world context. In *The origin of the Australians*, R. L. Kirk & A. G. Thorne (eds), 113–26. Canberra: Australian Institute of Aboriginal Studies.

Mayer, E. 1959. Isolation as an evolutionary factor. *Proceedings of the American Philosophical Society* 103, 221–9.

Maynard, L. 1979. The archaeology of Australian Aboriginal art. In *Exploring the visual art of Oceania*, S. M. Mead (ed.), 83–110. Honolulu: University Press of Hawaii.

Nanson, G. C., R. W. Young & E. D. Stockton 1987. Chronology and palaeoenvironment of the Cranebrook Terrace (near Sydney) containing artefacts more than 40 000 years old. *Archaeology in Oceania* 22, 72–8.

Pardoe, C. 1984. Prehistoric human morphological variation in Australia. Unpublished PhD dissertation, Department of Prehistory, Australian National University, Canberra.

Pardoe, C. 1988. The cemetery as symbol. The distribution of prehistoric Aboriginal burial grounds in southeastern Australia. *Archaeology in Oceania* 23, 1–16.

Pearce, R. H. & M. Barbetti 1981. A 38 000-year-old archaeological site at Upper Swan, Western Australia. *Archaeology in Oceania* 16, 173–8.

Peterson, N. (ed.) 1976. *Tribes and boundaries in Australia.* Canberra: Australian Institute of Aboriginal Studies.

Peterson, N. (in collaboration with Jeremy Long) 1986. *Australian territorial organisation: a band perspective.* Oceania Monograph 30.

Santa Luca, A. P. 1980. *The Ngandong fossil hominids: a comparative study of a Far Eastern Homo erectus group.* Yale University Publications in Anthropology 78.

Smith, M. A. 1989. The case for a resident human population in the Central Australian ranges during full glacial aridity. *Archaeology in Oceania* 24, 93–105.

Smith, F. H., J. F. Simek & M. S. Harrill 1979. Geographical variation in supraorbital [torus] reduction. In *The human revolution: behavioural and biological perspectives on the origins of modern humans*, P. Mellars & C. Stringer (eds), 172–93. Edinburgh: Edinburgh University Press.

Stringer, C. B. 1984. The definition of *Homo erectus* and the existence of the species in Africa and Europe. *Courier Forschungsinstitut Senckenberg* 69, 131–44.

Stringer, C. B. & P. Andrews 1988. Genetic and fossil evidence for the origin of modern humans. *Science* 239, 1263–8.

Stringer, C. B., J. J. Hublin & B. Vandermeersch 1984. The origins of anatomically modern humans in western Europe. In *The origin of modern humans*, F. H. Smith & F. Spencer (eds), 51–135. New York: Liss.

Thorne, A. G. 1971a. Mungo and Kow Swamp morphological variation in Pleistocene Australia. *Mankind* 8, 85–9.

Thorne, A. G. 1971b. The racial affinities and origins of the Australian Aborigines. In *Aboriginal man and environment in Australia*, D. J. Mulvaney & J. Golson (eds), 316–25. Canberra: Australian National University Press.

Thorne, A. G. 1975. Kow Swamp and Lake Mungo. Unpublished PhD dissertation, Department of Anthropology, University of Sydney.

Thorne, A. G. 1976. Morphological contrasts in Pleistocene Australians. In *The Origin of the Australians*, R. L. Kirk & A. G. Thorne (eds), 95–112. Canberra: Australian Institute of Aboriginal Studies.

Thorne, A. G. 1977. Separation or reconciliation? Biological clues to the development of Australian society. In *Sunda and Sahul: prehistoric studies in Southeast Asia,*

Melanesia and Australia, J. Allen, J. Golson & R. Jones (eds), 187–204. London: Academic Press.

Thorne, A. G. 1984. Australia's human origins – how many sources? *American Journal of Physical Anthropology* **63**, 227 (abstract).

Thorne, A. G. & P. G. Macumber 1972. Discoveries of late Pleistocene man at Kow Swamp, Australia. *Nature* **238**, 316–19.

Thorne, A. G. & S. R. Wilson 1977. Pleistocene and recent Australians: a multivariate comparison. *Journal of Human Evolution* **6**, 393–402.

Thorne, A. G. & M. H. Wolpoff 1981. Regional continuity in Australasian Pleistocene hominid evolution. *American Journal of Physical Anthropology* **55**, 337–41.

Veth, P. 1989. Islands in the interior: a model for the colonization of Australia's arid zone. *Archaeology in Oceania* **24**, 81–92.

Webb, S. G. 1984. Intensification, population and social change in southeastern Australia: the skeletal evidence. *Aboriginal History* **8**, 154–72.

Webb, S. G. 1989. *The Willandra Lakes hominids*. Canberra: Department of Prehistory, Research School of Pacific Studies, Australian National University.

Webster, P. J. & N. A. Streten 1978. Late Quaternary Ice Age climates of tropical Australia: interpretations and reconstructions. *Quaternary Research* **10**, 279–309.

Weidenreich, F. 1943. The skull of *Sinanthropus pekinensis*: a comparative study of a primitive hominid skull. *Palaeontologia Sinica* D, **10**.

Weidenreich, F. 1947. Facts and speculations concerning the origin of *Homo sapiens*. *American Anthropologist* **49**, 187–203.

Weidenreich, F. 1951. Morphology of Solo man. *Anthropology Papers of the American Museum of Natural History* **43**, 205–90.

White, J. P. & P. J. Habgood 1985. La préhistoire de l'Australie. *La Recherche* **167**, 730–7.

White, J. P. & J. F. O'Connell 1979. Australian prehistory: new aspects of antiquity. *Science* **203**, 21–8.

White, J. P. & J. F. O'Connell 1982. *A prehistory of Australia, New Guinea and Sahul*. Sydney: Academic Press.

White, N. G. 1979. The use of digital dermatoglyphics in assessing population relationships in Aboriginal Australia. *Birth Defects: Original Article Series* **15**, 437–54.

Wolpoff, M. H. 1980. *Paleoanthropology*. New York: Alfred A. Knopf.

Wolpoff, M. H. 1985. Human evolution at the peripheries: the pattern at the eastern edge. In *Human evolution, past, present and future*, P. V. Tobias (ed.), 355–65. New York: Liss.

Wolpoff, M. H., Wu Xin Zhi & A. G. Thorne 1984. Modern *Homo sapiens* origins: a general theory of hominid evolution involving the fossil evidence from East Asia. In *The origins of modern humans: a world survey of the fossil evidence*, F. H. Smith & F. Spencer (eds), 411–83. New York: Liss.

Yamaguchi, B. 1967. *A comparative osteological study of the Ainu and the Australian Aborigines*. Australian Institute of Aboriginal Studies, Occasional Papers 10, Human Biology Series 2.

Index